Things My F' ˙ ˉ
Nevei

Other Lessons Learned in
25 Years of Flying.

By Michael Leighton

Published by
Promotion Productions Inc.
P.O. Box 740020
Boynton Beach, Florida, 33474-0020
Phone: 561-752-3261
e-mail: proprodinc@aol.com

ISBN: 0-9760898-0-7

Printed in U.S.A.

Things My Flight Instructor Never Told Me

Introduction I
#101-How I Avoided Becoming an Aviation Statistic 3
#103-Embedded CB's 7
#105-Hard IFR 13
#106-Operations at an Out of Control Airport 20
#107-The Instrument Proficiency Check 30
#109-The Biannual Flight Review 35
#110-Out of Control 39
#111-Belt and Suspenders Approach 45

Other Lessons Learned
The Organized Pilot 51
A View From the Right Seat 56
Bad Service 62
Some Thoughts on Density Altitude 72
Pushing your Personal Envelope 76
Watching Good Training in Action 81

Adventures in Aircraft Ownership
Lien on Me 84
No Damage History 92
Witness to Stupidity 98
Things You Can Only Learn in a Piper Cub 104
Stepping up to a Twin 122
The Realities of Flying a Multiengine Aircraft 131
So, you Want to Fly a King Air? 156
Flying the Citation Jet 163
Tail Wheel Tips 174
The More Things Change 178
About the Author 187

Things My Flight Instructor Never Told Me

and

Other Lessons Learned in 25 Years of Flying.

By Michael Leighton

Introduction

My business partner, a 20,000 hour former Marine Corp fighter pilot describes experience as "what you get when you don't get what you want." As it applies to aviation, I could not agree with him more.

In the 25 years that have transpired since I took my first flight lesson, I have received my fair share of "experience".

Along the way, I started writing about the things I learned, the planes I flew and the people I have met.

Things My Flight Instructor Never Told Me is a collection of stories chronicling my experiences as an airman, and as an aircraft owner.

I want to say a special thanks to a mentor and a friend, whose seemingly endless love and knowledge of all things aviation has not been lost on me. For without his insights, guidance and occasional helping hand, I would have walked away from aviation a long time ago.

Thanks Owen, I would not have done it without you.

Things My Flight Instructor Never Told Me #101

Or

How I Avoided Becoming Another Aviation Statistic

Iroutinely tell this story to my flight students to make a point. This particular story would be humorous if the consequences weren't so severe.

In the early 1990's, I had a business trip to do from Palm Beach, Florida to Columbus, Georgia, spend the night and return the next morning. About 400 nautical miles one-way, it was well within the range of the Cherokee 140 I was considering doing it in. I like the Cherokee because they are such simple, easy to fly forgiving machines that it doesn't take long for me to develop confidence in the particular airplane. I had flown this one for about 15 or 20 hours since I had bought it, checked out all the radios and they were good. This one had two unusual pieces of equipment in it that I was interested in trying out on a cross country. One was a Collins DCE. This is a device that takes two VOR signals, and essentially becomes a poor man's DME, giving you ground speed and distance information. The other was an S-Tec 50 autopilot. This is a lot of autopilot for a Cherokee 140 but the previous owner had installed it and I was interested in trying it out on a long cross country.

So I loaded up and took off IFR on what should have been nothing if not a routine flight. The climb to 8,000 feet took awhile but once there the ride was smooth (for south Florida in spring time) and I was in and out of light build ups for the first hour or so. I did all the house keeping tasks including verifying the 104 knot ground speed my DCE was showing, switched fuel tanks, slid the seat back to have some breakfast and watch the autopilot fly. I had never had an autopilot with altitude hold before and I was enjoying it. I turned my attention to fuel management. This is about the only thing you can do wrong in a Cherokee.

I took off with 50 gallons. I knew that because I filled the tanks myself. The book says fuel burn at that altitude at 2400 rpm should be about 6.5 gallons per hour. Experience told me more like 8 gallons per hour. I always like to be conservative with fuel burn. Plus you always burn more the first hour because of the time in the climb at high power and usually a full rich mixture, so I figured 10 gallons per hour the first hour, 8 gallons per hour every hour after that. That should give me close to six hours endurance at that power setting. (8 x 4 is 32, plus 10 is 42 gallons at the five hour mark plus 8 gallons in the tank or 1 hour reserve, my personal minimum).

I ate my croissant and drank a cup of tea and watched as the autopilot tracked the center of the airway with a precision I could never duplicate by hand.

At the three-hour mark I began to believe that the tea was probably not a great idea. I had enjoyed one other in flight experience with the so called "pilots pal" and it was let us say disastrous, so I was apprehensive at best when I realized that I must either reach for the jar, or land.

I opted for the jar. That was my second mistake. While cursing my stupidity to myself and chastising myself for making the same mistake twice (I'm much more forgiving of new mistakes, but the same one twice really ticks me off) I contemplated the practical realities of installing a relief tube in my airplanes from now on.

I turned my attention back to flying to get my mind off my "situation". The flight was pretty much in the clear, it was VFR everywhere, and cancelling and landing somewhere would not be the monumental hassle it would be if I had been solid IMC and begging for an amended clearance. But the engine just purred, everything was in the green at the four hour mark I was 60 miles from my destination. About the only thing that bugged me at all was that the fuel gauges were beginning to bounce off the "E" side of the scale. Anyone who has flown a Cherokee knows that the electric gauges in the panel are hardly worth the space they consume. They are about as believable as a witness to an alien abduction. I always fly all my airplanes the time honored way. Know what you started with, and use a watch. I re-figured the fuel burn, rechecked the book and even figured that if the gauges were right I'd have to have been burning about 12.5-13 GALLONS PER HOUR......in an Lycoming 0-320!

But right about then nature's call returned. Not willing to make the same mistake 3 times.....twice in the same day, I elected to cancel and land at Albany, Georgia, five miles off my nose.

The landing was uneventful. I hopped from the plane asking the line crew to top it off and I headed for the restroom. When I returned the lineman handed me a fuel receipt for 48 1/2 gallons.

I was stunned. I would never have made it to Columbus. Instead of landing with 8 gallons in the tank I would have parked it in a field short of my destination. I popped the cowl and looked under the wings for a fuel leak, anything that could explain how I could get that close to committing one of the cardinal sins of aviation.....fuel exhaustion!

That night in Columbus, I lay awake in bed for hours replaying the flight, but nothing came to mind.

The next day on the return trip, I planned to go to Sanford, Florida, north of Orlando, stop for fuel, re-check the fuel burn per hour, eat lunch and fly the rest of the way home.

Before leaving Columbus, I was sure the tanks were absolutely full to the rim. 3 1/2 hours later I landed in Sanford, and took 39 gallons. That worked out to 13 gallons per hour in a Cherokee 140, twice the book number.....explain that!

Three days later my mechanic did. Apparently, on this particular engine installation there is a vent line that runs from the top of the carburetor float bowl back to the fuel tank to carry overflow from the float bowl back to the tank. This line was cracked, and instead of returning the fuel to the tank, it was venting it to the atmosphere. The line is in the high-pressure part of the cowl, and the warm fast moving air simply atomized the fuel before it had the opportunity to stain anything blue, thus no telltale signs of the leak. The higher I flew the worse it made it because of the lower ambient air pressure relative to what was going on inside the float bowl.

If I hadn't had that cup of tea, if nature's knock had come a little later, I would have become a footnote on the bottom of an accident report.............. fuel exhaustion due to pilot error. The fact is the aircraft didn't burn 13 GALLONS PER HOUR, but that is what I was losing between what the engine needed to run and what it pumped overboard. I didn't think it was possible, but it obviously is, and that is **something my flight instructor never told me.**

Things My Flight Instructor Never Told Me #103

EMBEDDED CB'S

Flight instruction is such a complicated thing. To do it properly you really need to teach from a position of experience. That is exactly the opposite of the majority of the flight training industry. Experienced pilots don't necessarily make good teachers, but in general, flight instruction is perceived as a steppingstone along the path to a "better" flying job. Most who pursue that path are not really interested in becoming a good flight instructor, they are interested in logging enough flight time to get on with an airline. Predictably, the quality of the education that the new pilot receives is sometimes less than complete. But because an aspiring pilot simply doesn't know what they don't know, he or she accepts whatever level of flight training they receive as adequate. Indeed, that must be so or they wouldn't pass the check ride! (Right?)

But early in my flight training, I sensed that the young guys teaching me had their attentions elsewhere and I sought out the older wiser instructors to fly with me.

I had received my instrument rating in 1988 from a name brand Academy type school, and taken additional practical and simulator training from an airline pilot who operated a flight school on the airport in New Jersey that I had learned to fly at years earlier.

Watching that old airline pilot fly was a treat. He would set the simulator, an AST 300, to level five turbulence, dial in a forty knot direct crosswind and shoot a perfect approach. His hands and feet would never stop moving, but the artificial horizon would remain rock steady, and the plotter would scribe a perfect approach, every time.

He was a tough instructor, pushing you to your limits and beyond every time you gave him a chance. I would go up to his school whenever I was in the area on business and grab some time in the simulator. That training probably saved my life.

That sense of security became a double-edged sword one June afternoon in 1990. I had finally bought an airplane that was capable of serious cross-country flight and I was enjoying it. The 1977 Mooney M-20J was pretty well equipped for a light plane of its time. It featured dual Nav-Coms, ADF, DME, audio panel with intercom, and a Century 2 autopilot. It was a step up performance wise from the Cessna Cutlass I had gotten my instrument ticket in.

I had owned the plane for several months and received plenty of dual from another very experienced instructor I had befriended when I moved to Florida. That man, a twenty thousand hour gold seal instructor had a very casual, matter of fact way about him. The exact opposite of the airline pilot I had trained with in New Jersey.

Feeling good about my airplane, and myself I decided it was time to take the Mooney cross-country. My wife and I headed off with another couple from our base in West Palm Beach, Florida toward Morristown, New Jersey. The weather was typical east coast in the early summer. Haze and low flight visibilities, surface temps in the 90's along the entire route of flight and of course, the possibility of thunderstorms. In Florida, if you don't fly anytime there is a chance of thunderstorms in the forecast, you may only be able to fly 25 days a year. But here in Florida, thunderstorms are easy to see and avoid, and typical flight visibilities even in the summer are better than 10 miles. I completed my preflight, and got an updated weather briefing. The trip, from West Palm Beach to Morristown, New Jersey with a stop in Florence, South Carolina would take about seven hours.

The first leg to Florence went well. Early morning skies over Florida were clear and as we crossed into Georgia we began to see the haze. The visibility at Florence was a questionable five miles, but clear of clouds. While we fueled I checked the weather. The radar was showing a small line developing west of Victor 3 on a roughly northwest southeast line. It was moving east at about 20 miles per hour.

I elected to amend our flight plan to take us up Victor 1, which runs parallel to Victor 3 but is about 80 miles east on average. The destination weather was good and once north of Richmond, Virginia area, it appeared as though we should expect a nice trip.

As we taxied out, we were advised of a new sigmet, which had just been issued for our route of flight. After takeoff, I called flight watch for an update on that sigmet. The line of thunderstorms I had viewed on the radar less that 20 minutes earlier had turned nasty. We had filed for 7,000 feet. As we approached the Richmond area, we began to hear aircraft asking for deviations from the weather.

Again, I called flight watch to check on the situation, just as I had been taught. Flight visibilities were now nil, and we were solid in it at 7,000 feet. Flight watch reported that the line was moving as before but that we were at least five miles east of the nearest echo, which was the southeastern tip of the line, stretching to the northwest. Switching back on the center frequency we could now hear regional airliners frantically requesting course deviations for weather. Less than a minute later it got really dark in the cockpit. The sky turned that grayish green color that you know is associated with nasty weather. The static electricity in the air was so intense that everyone's hair started to stand up as if we had placed our hands on a Van de Graff generator. Then I saw the first lightning flash. I turned to my passengers and told them to tighten their seat belts. I looked at my right seat passenger, a private, non-instrument rated pilot and reminded him not to touch the controls. Then we hit the updraft. I'll never know how strong it really was because the VSI was pinned in the 3000 fpm up position. I turned off the autopilot, closed the throttle, lowered the landing gear and put the nose down and trimmed to maneuvering speed. Still going up. As we went through 9,000 feet, center asked me to say my altitude. I told them 9,000 and climbing. He asked if I had been assigned 7,000. I answered affirmative but that I was in a cell right now and was just trying to keep the wings level. He cleared me to deviate as required and suggested a heading of 088°. Just as he gave us that heading, we hit the down draft. The VSI swung around in the other direction. Full power, gear up, nose up to Vy plus or minus and still going down.....in excess of 3,000 fpm! At that point we penetrated the rain shaft. The noise was unbelievable. The turbulence racked the plane left and right. As I attempted to turn to the right toward the assigned heading the plane went up on the right wing at close to 90° of bank. Full left aileron and rudder took nearly five seconds to recover to wings level. Stuff was flying around the cockpit. The wife of my right seater and my wife were in the back seat.

Going through 5,000 feet I remember thinking, "I hope we run out of this storm before we run out of altitude". I was amazingly calm, almost detached from the fray. I though to myself, "so this is how people die in a small airplane". I continued to work on keeping the wings level and to set the pitch with the airspeed indicator. It seemed like two hours but it was more like two minutes. All of a sudden, we popped out of the side of the towering Cu, into a clear blue cloudless sky. The feeling of relief lasted only a second or two as a horrendous sound filled the aircraft. It was the sound of 2 to 3 inch hail striking the airframe at nearly 200 mph. We had penetrated the cell from behind and came out the front, under the anvil and the falling hail. We had survived the encounter with the storm only to be assaulted in clear air by ice balls the size of golf balls.

The remainder of the flight into Morristown, New Jersey was uneventful. In fact, it was a rather nice day in the northeast, with flight visibilities in excess of 15 miles and few clouds at any altitude.

The full extent of the damage to the airplane didn't become evident until we deplaned. The entire leading edge of both wings, the tail and top of the rudder looked as if a madman had attacked us with a ball peen hammer. The hail strikes were everywhere. Blue fuel stains on the bottom of the wing panels indicated we had ruptured every fuel tank seam on the airplane. It was a miracle we didn't lose the windshield.

That night I lay in bed replaying the day over and over in my mind. I did everything right, everything I had been trained to do. So how did I get so close to becoming a footnote on an NTSB report in the first place?

Nowhere in any of my instrument training had anyone ever mentioned that you should never, ever fly in IMC in embedded thunderstorms. That thunderstorm avoidance is strictly a visual thing. And thunderstorm avoidance gear such as Stormscope™ and radar (which was not that common in singles in 1990) are absolutely mandatory and still not fool proof for operating in those conditions.

What I should have done, and now do, is operate at an altitude that puts me above the haze so I can see the build-ups. I fly around what I can't fly over. This often means operating at ten to twelve thousand feet or higher in the summertime. All my aircraft have been equipped with storm detection equipment of one sort or another and I carry a small portable oxygen bottle.

The cell I encountered was certainly not larger then a level one. I have no interest in seeing what a level 2 is like. I now make it a point when giving instrument dual to take my students on a crosscountry trip that puts them in this situation so they can see for themselves what you can and can't safely fly though.

Even fast building Cus that haven't begun to spark yet can be dangerous.

The bottom line; **DO NOT FLY IN INSTRUMENT METEORLOGICAL CONDITIONS WHEN EMBEDDED THUNDERSTORMS ARE PRESENT OR FORECAST!** Something that my flight instructor never told me.

Things My Flight Instructor
Never Told Me #105

HARD IFR

Experience is by far the best teacher. My partner says experience is what you get when you don't get what you want. When it comes to learning things aviation, I have found that it is always more comfortable to venture into the territory beyond your own personal experience, while in the company of someone who has been there and done that. This way, if things do go to hell in a hand basket you have the voice of experience to listen to. But if you fly long enough the day will come when you know a lesson is coming and you are going to have to go out and take it alone.

So it was for me on the morning of March 4th. Our charter company had a trip scheduled for some time that was in direct conflict with a training program we had scheduled for our new hire captains. Being the lowest time captain in our group it was decided that I would forgo the training program (which was for another aircraft that I had no intention of flying anyway) and fly this particular trip, so that the new hires could get qualified on the new aircraft.

That morning the weather map was a classic late winter picture. A cold front extending from over the Great Lakes into the Gulf of Mexico was dumping snow all over the Midwest. A warm front, created from the remnants of the last cold front had backed up over the state of Florida moving north, attached to a low moving northeast up the cold front.

Our trip was from our base in South Florida, up to Jacksonville, to pick up our client, then on to Teterboro, New Jersey for the night. The next morning we were to go on to

Bradley International in Hartford, pick up three more passengers then return to Jacksonville, and on home to Palm Beach.

The northbound trip was a cinch, featuring clear skies and great visibilities not to mention a 25 kts tailwind going north ahead of the approaching cold front. Other than the usual crummy service from Philadelphia and New York approach, the leg went off without a hitch. Our client landed within five minutes of our projected ETA even with all the rerouting and the scenic guided tour of the eastern Pennsylvania / New Jersey border. After putting the plane to bed and catching some dinner, my copilot and I sat down to look at tomorrow.

The weather in the morning in the northeast would be fine. The front would not make its presence felt here until late afternoon, and we would be a thousand miles away by then. But as the warm front, now across northern Georgia moved north, a trough emanating from the low over Louisiana began to move north and east. The same features that gave us those welcomed tailwinds would now create strong headwinds, and it was likely we would not be able to make the trip from Hartford to Jacksonville non-stop.

But the die was cast. We were supposed to be professional pilots. The chief pilot didn't want to hear about weather. The customer didn't want to hear about weather, and I was quite sure the three passengers we were picking up in Hartford didn't want to hear it either.

Fortunately, my customer is a pilot and a close friend as well and does not suffer from "getthereitis" at all. So I always had the "wait till tomorrow" card to play if the weather exceeded my personal comfort level. Knowing this I went to bed early because one way or another tomorrow was going to be a long day.

Up at 5a.m. I jumped on the computer to get the forecast. We had recently invested in a few new laptops and my partner loaded them up with all sorts of cool flight planning software. For those of you not using computer flight planning, you don't know what you are missing. Not only does it have all the current airways, both high and low altitude, but the regular update package which comes on CD, has all the current procedures in it for route decisions. If you are not familiar with the area and how ATC works there, use the router feature and it gives you a route that on most days you will actually get. I'm sure most of the flight planning software does that but hey, in the middle of the busiest airspace in the country, to hear "cleared as filed" is nice.

Then it calls the weather for you. It will connect to DUAT or Dynacorp it doesn't care, just tell it which. It will overlay the weather graphics right onto your route map so you can see where the weather is and where it is forecast to be. It will even get the current TFR's. You can read it in plain English or if you insist, in code. But the feature that paid for itself on this trip was the "Wind Optimize" feature. Tell the computer what you want to do and it will tell you which altitude will give you best speed vs best fuel burn. No matter how I sliced it, we were not making it non-stop without going right to our one-hour fuel reserve. That fact, coupled with the forecast for weather at the destination had me scrambling to re-file with a fuel stop at Salisbury, Maryland. The weather there was forecast to be VFR throughout the morning and I figured we would stop, fill up and get one more weather briefing before taking on the serious weather.

The first leg up to Hartford and the second down to Salisbury went off without a hitch. But it was apparent that the winds were definitely stronger than forecast and our ground speed confirmed that.

At Salisbury we got on the phone with the briefer while the plane was being fueled.

Airmets for turbulence at almost all altitudes, Sigmets for ice and low ceilings and visibilities along the route was what we heard and what we expected. Destination weather was still o.k. and forecast to be 1,500 and three on arrival. It was decision time.   Upper level winds were not only stronger than forecast but the higher we went the less favorable the direction, until above FL 200, where they were right on the nose. The flip side to that was the higher we went the lower the temps, thus a lower chance of severe icing, plus better fuel burns. At 22,000 feet, we would arrive over the destination with more than two hours of fuel, and the weather got better further south of Jacksonville. We decided to go.

On the way to FL220 we were in the clear but a hundred miles down the road we were solid IMC and minus 22°C. It stayed that way until we broke out on the approach nearly four hours later.

All around us we heard airliners asking for higher or lower to get out of the ice. We didn't see any ice until well south of Norfolk. Then we began accreting ice on and off until from about 100 miles north of Charlotte we were picking up ice fast enough so that I had to cycle the boots every few minutes. The turbulence was intermittent and never more than occasionally moderate. The big feature was the wind.

We were cleared direct to Craig Field from 100 miles north of Charlotte. The ground speed was down to 130 knots, a hundred knots right on the nose. The time to station on the GPS read 2:48. It stayed there for nearly an hour. At one point, my customer leaned into the cockpit and inquired if the damn thing was broken because it had the same time to station readout from an hour ago. I told him no, it worked fine; we were just going so fast that we were going back in time. He laughed and looked forlornly at the potty, which was covered with golf clubs from the other three passengers.

Approaching Savannah, the controllers advised us of convective sigmet Echo 21 for a line of thunderstorms from 30 west of Jacksonville, to Ft. Meyers, twenty miles wide moving northeast at twenty. The time to station display on the GPS read 1 hour. We were now in a horse race. I figured if it was 30 west, moving east at 20mph, we had about an hour. I also knew we would pick up some time on the long descent from FL 220. If we didn't get vectored half way to Disney World, we might make it.

South of Savannah, Jacksonville was reporting better than 5,000 and five. The radar was showing a few small cells, which we could easily deviate around. We were told to expect the visual to runway 14 which was kinda in the direction we were going....ahhh the best laid plans.... Coming through 8,000 feet we could see the coast through the undercast and the line of thunderstorms appearing on the edge of our airborne radars range. A last minute vector south and east of the field to put us in trail behind a Skyhawk was an unwelcome surprise. "What was a student and an instructor doing out on a day like this anyway", I thought to myself. Either they wanted some actual, or they got caught out and were trying to get home. Either way, it sent us 15 miles out of our way. As we turned base to final the radar swept to the west revealing the imminent deluge lurking in the gloom just west of the field. As we taxied in, a lightning bolt struck a tower on the airport leaving a loud report for my passengers to contemplate. As they deplaned, the first large super-cooled water droplets began to fall on the ramp as I sought refuge under the tail. With the passengers safely in their rental car, I closed the cabin door and beat a path to the FBO. It poured for more than an hour.

My copilot and I sat in the crew lounge, ate our lunch and watched the weather on the five o'clock news. Our desire was to make it home to Palm Beach and sleep in our own beds. The weather would have the final say.

An hour later, the first line had moved through and we had a window of opportunity to get out of Dodge and head south. Weather along the route was fine inland, nasty on the coast. It was dark by now so we filed and fired up. On departure the controller offered us vectors for weather without us having to ask. Jacksonville always has great service and tonight it was really appreciated. By now we had been up 13 hours, and flying more than eight. We were tired but the night air was smooth as glass the plane was running great and there was a real prospect of VFR on arrival. South of Orlando we broke out into severe clear and all of a sudden even the persistent headwind didn't seem so bad. We touched down 12 hours to the minute from departure at Teterboro. On any "normal" day that was a six-hour deal. Today, thanks to a late winter coastal storm it was double that.

The moral of the story? My partner was right. Experience is what you get when you don't get what you want. If I had been in a lesser aircraft and not a twin-engine turboprop with all of its high altitude and known ice capabilities we would not have gone. Our decision making process was sound and based on aircraft as well as pilot capability. Having the "out" plan ready if the destination weather went to hell prior to arrival as well as the command authority to "just say no" in the first place all played a role in the flight planning. There is nothing that I can think of in my flight-training syllabus that could have prepared me for this decision train. **Only experience and confidence in my equipment and my personal flight skills could have guided me in evaluating this particular go-no go decision. That is something my flight instructor never told me.**

Things my Flight Instructor
Never Told Me #106

Operations at an Out of Control Airport

At first glance, the general aviation airport that I have called home for the last 13 years is a typical small airplane airport. Jets are banned here, and there are multiple flight schools, as well as helicopter and banner towing operations mixed in with the occasional turbo-prop and ultralight. Even a blimp calls our little airport home several weeks a year, mostly during football season.

But for all of the activity, our airport is an uncontrolled airport. Some say it is an out of control airport. Were you to fly in here on any winter weekend afternoon, you would have to agree. Some of the congestion comes from the fact that we underlie Class C airspace, and are indeed only 5.5 nautical miles from the Class C airport. There is a cutout in the inner ring of the Class C to accommodate our traffic area, but it forces most of the traffic to approach from the south. Aircraft arriving from the north are talking to approach, and are kept up at 2,500 feet or higher until south of the primary class C airport, so you are forced to descend into the pattern coming from the north.

The unfortunate reality is that there are no regulations as such governing operations at an uncontrolled airport. If you look at the AIM, part 4-3-3 discusses recommended procedures for ops at uncontrolled fields, but the only regulation cited in all of that text refers to FAR 91.113. That particular regulation simply states that the lower aircraft has the right of way, but shall not take advantage of that rule to cut in front of another aircraft on final approach or to overtake that aircraft.

In fact, if you really want to start a fight, just go ahead and point out to the bonehead who just made a right turn on departure that there is a left hand traffic pattern, and watch the fur fly.

I have had nose to nose shouting matches with instructors who smugly point out that there is no actual regulation requiring them to fly the left hand pattern, so therefore, I should shut up and mind my own business. They use the same argument when it comes to straight in approaches, or entering the pattern on the 45° (or the crosswind) as opposed to the midfield overhead break.

These are usually the same pilots who never shut up on the radio, making position reports starting at 10 miles out about every 2 miles, then crosswind, downwind, mid-field downwind, base, final, short final and clear of the runway.

In between position reports they can be heard calling other traffic to verify their positions, and then advising them whether or not they are "no factor" in their traffic pattern. Let me also mention that on many occasions the size of the traffic pattern can get so stretched out that I you can routinely see aircraft turning final three miles from the airport. It is also common to see more than one runway in use at a time, and there is no regulation that would preclude that either.

The problem with operation in a manner other than what is recommended in the AIM is not regulatory; it is safety that is being compromised.

Before you start typing those poison e-mails go look at the accident statistics concerning mid-air collisions. Most mid-airs occur at uncontrolled airports on clear days. Most mid-airs occur between two aircraft on board one of which is an instructor. A good percentage of these accidents occur between high wing and low wing aircraft. What I am getting to is this; if you operate at an airport like the one I described, you have a higher than normal chance of being involved in a mid- air. So why wouldn't you do absolutely everything possible to reduce your risk? Just because there is no rule that says "you must" doesn't mean, "you shouldn't".

So lets discuss the things we can do to protect ourselves from unnecessary risk, while utilizing proper pattern etiquette.

First, all of VFR is based on the concept of see and avoid. Nowhere is it implied that you should talk and avoid. Way too much emphasis is put on radio transmission of positions while in the pattern, and not enough on keeping your head outside where it belongs, actually <u>looking</u> for the traffic. On that subject, keep your transmissions as short as possible. Listen first. Nothing is a greater waste of airtime than to hear three requests for airport and runway advisories in a row in a span of 25 seconds. If the other eight aircraft in the pattern are all using runway three, it is good bet that runway three is the active runway. Do you really need to call the Unicom operator and have them tell you that personally? Identify yourself as Cessna or Twin Cessna and the last three numbers of your tail number.

Yes, you can call yourself Skylane, Skyhawk or Stationair but to all the low wing drivers out there in the pattern with you, they all look the same.

Next, listen to the Unicom ten miles out and figure out which runway they are using. Don't make your first call till you are no more than five miles out. If you're flying a high performance aircraft, slow down. There is absolutely no point in blowing into the pattern at two hundred knots simply because you can, when the traffic pattern is limited by the slowest airplane in the pattern at the moment. If you are flying the slowest ship in the pattern realize that some of the high performance aircraft fly the pattern at a speed faster then your cruise speed, so be considerate and don't maneuver to cut that Aerostar off on the downwind, because he can't go any slower. His only option is to fly a large pattern to give you the time to land.

Twelve years ago I witnessed four people die in a fiery crash in the traffic pattern at our airport. The pressurized piston twin was on its third attempt to land after being cut off on final during the previous two attempts by two different training aircraft. On the third attempt, it happened again, this time by a trainer waiting patiently at the departure end waiting for several minutes, that had pulled out in front of the landing twin. The pilot of the twin lost his cool, shoving the throttles forward while the aircraft was low and slow with gear down and full flaps deployed. He had dropped below blue-line anticipating a landing when he executed the go around. The aircraft started a Vmc roll over and the pilot had partially recovered when the plane pancaked into a stand of trees in a pond on the west end of our field. My buddy and I were the first ones to get there. The two women in the back seat were screaming, the guys in the front seats were motionless. As we approached the right wing, the fuel from the ruptured fuel tank which was floating on the pond, itself only a foot or so deep, ignited and we were forced to retreat. The plane exploded, and everyone stopped screaming.

Several years later, an experienced former airline captain returning to our field with an engine out on a twin Cessna, died in a Vmc roll over accident right in the middle of the field. To make things worse, the local news crew was out there taping something or another that day and the whole horrible mess showed up on the six o'clock news on every station for two days. The pilot had declared an emergency and was speaking to the other aircraft in the pattern. He had flown a large pattern so as not to have to turn into the dead engine or bank too steeply, when a student with an instructor on board in a trainer cut him off in the pattern and landed on the runway ahead of him. For some reason the pilot and instructor in the trainer decided to use the whole runway to roll out on and the pilot of the twin was concerned that there might not be room for two aircraft on the same runway. He attempted a go around, without success. At the same airport, a Twin Otter with about a dozen people on it departed that same runway. It returned a few minutes later with one engine feathered. The pilot declared an emergency, and we heard it on the Unicom frequency. Just as he had the runway made, an aircraft pulled out on the active runway to takeoff. The Otter pilot attempted a single engine go around. The aircraft had other ideas. The airplane came to rest off the departure end of the runway in a stand of trees. Miraculously no one was seriously hurt and the pilot was heralded a hero. The pilot of the aircraft that had pulled out in front of the stricken twin claimed to have never seen or heard the emergent aircraft.

The point is, make the effort to enter the pattern and fly the pattern the way it is depicted in the AIM. The reasons for this are numerous but consider the following; Entering the pattern on the 45° to the downwind or on the crosswind allows you, the pilot, sitting on the left side to watch the traffic coming off the runway the entire time without losing sight of it. (In a left hand pattern.) It also lets everyone else know exactly where to look for you. The 45 joins the downwind at mid-field but you are supposed to use that as the variable to allow you to fit yourself into the pattern.

Never shoot a straight in approach. Doing so might save you two whole minutes but what if someone is using one of the other runways, or the opposing runway? If you fly the pattern, you will see him if he doesn't see you.

Always turn in the direction of the traffic pattern. If you turn right on departure because you want to go that way, in a left hand pattern, you are turning directly into the crosswind approach to the airport. If you just have to turn against the pattern, at least fly a straight out departure far enough from the airport that you are well outside of the traffic pattern.

The best and last piece of advise I want to impart here is this; If you arrive at the airport and the fur is flying, the pattern looks like opening day at Oshkosh, you can't get a word in edgewise and you not sure if you want to.....just fly away. Come back in five minutes and it will all be gone. Do not try and fit yourself into the melee. Avoid the mindset that somehow you have to make this mess work and just fly away. **That little piece of advice is something my flight instructor never told me.**

Don't even try to figure out what the other guy is doing. Remain in control of the situation by remaining in control of your aircraft. If you get cut off on final, move to the right (runway on your left) so you can watch the offending fool as you execute your missed approach. If he just cut you off on final, so who knows what else he is capable of doing.

Not every uncontrolled field is going to be this exciting, but if you fly long enough, you're going to find yourself in a similar situation. Fly the pattern by the book, reduce your risk of a mid-air and show your best pattern etiquette.

91.126 Revisited

When I wrote the story about operations at uncontrolled airports, I knew that it was a controversial subject. While most of you are familiar with FAR 91.126, I thought it would be worth another look. So lets take a look at 91.126 and see just exactly what it does and doesn't permit.

Sec. 91.126 Operating on or in the vicinity of an airport in class G airspace. (a) *General.* Unless otherwise authorized or required, each person operating an aircraft on or in the vicinity of an airport in a Class G airspace area must comply with the requirements of this section. (b) Direction of turns. When approaching to land at an airport without an operating control tower in Class G airspace-

(1)Each pilot of an airplane must make all turns of that airplane to the left unless that airport displays approved light signals or visual markings indicating that turns should be made to the right, in which case the pilot must make all turns to the right and...

What is significant here is the language, and what it does not say. Note that it says all aircraft <u>approaching to land</u> must make all <u>turns</u> to the left. It <u>does not</u> say you must fly a left- hand traffic pattern. If you make no turns at all, that's o.k. . Straight in approaches are "legal".

Furthermore, it only refers to aircraft approaching to land. It does not specifically address aircraft departing. A right turn on departure is also "legal" as the aircraft is not approaching to land. Under 91.126, it is "legal" to enter the traffic pattern on the base leg so long as the base to final turn is to the left. Lastly, it directly contradicts the AIM on the subject of a midfield 45°entry, as that would require a right turn. So, like it or not, that is what the regulations say.

Traffic patterns per se, are not specified and are absolutely not regulatory.

At one point, I have had a very heated discussion with a very experienced instructor with more than 20,000 flight hours. These are the type of instructors that young airmen look to for guidance and try to emulate in their flying. This man simply did not agree with the concept of a left hand traffic pattern and insists that he will fly a right hand pattern, even at an airport with left traffic because he feels it is safer.   That might be so if everyone in the pattern had 20,000 flight hours but I perceived his response as advocating anarchy.

In the middle of that discussion, another experienced pilot put the blame for pattern confusion at uncontrolled airports on "out of control" pilots. He went on to build his case that many use the radio in lieu of a standard pattern for traffic separation. On this point few can argue. The concept of VFR traffic separation is based on see and avoid. Nowhere in the FAR's or AIM does the concept of talk and avoid come into play. Unfortunately, so many low time pilots have been taught to talk and not to look, that this type of behavior has become standard protocol at many uncontrolled G.A. airports.

Not long after I wrote that story, I was taking a part 135.293 / 297/ 299 check-ride with an F.A.A. examiner. We were shooting a VOR/ DME arc approach into an uncontrolled airport on a marginal day. It was VFR, but we were on an IFR flight plan. I made my first Unicom radio call 15 miles out. I made another at 10 miles out, followed by yet another at 5 miles out. Remember, we were IMC for real. There were 2 aircraft in the pattern that I could hear on the radio. We were approaching from the north and the landing runway was 9. It would be a circle to land. We broke out at 1,500 feet, 10 miles from the airport. I could clearly see the traffic in the pattern, so visibility was not that bad. I descended to the MDA, which is 660 feet, and leveled off, intending to fly to within one mile of the field (circling minimums) and join the pattern. I made a position report at five miles out another at two miles out. I was watching and listening to a Skyhawk doing touch and goes.   It was apparent it was a student and an instructor. They too were making position reports, about every 15 seconds. Upwind, crosswind, turning downwind, mid-field downwind, he just didn't shut up. In between, he was "chatting" with the instructor in the other aircraft also doing touch and goes. As I turned right to enter the downwind, below and behind the Skyhawk, the instructor came on the radio and asked if we were in the pattern. When I advised in the affirmative, he asked why I hadn't announced my position until I was directly under him!

The Fed in the airplane with me about came unglued. Keying the mic he ripped into this guy telling him that he should spend less time talking and more time flying. It got really quiet on the frequency. Feeling sorry for the guy, I shot the missed approach and headed for home before the Fed could insist that I land so he could "talk" with the guy in the Skyhawk some more.

The point is, how can you expect more from a low time or student pilots when the instructors are guilty of poor pattern performance? The answer is in each and every one of us. Use proper pattern procedures, and take the time to educate those unfamiliar with the idiosyncrasies of your particular field to new or transient pilots. Recognize that there are "holes" in the regulatory text that allow straight in and other "non-standard" traffic procedures and as pilot in command, you need to be prepared to deal with it.

TMFINTM #107 -

The Instrument Proficiency Check

Every instrument rated pilot is familiar with the regulations concerning currency. Once called the six, six in six rule, for six approaches, and six hours of flight in IMC in the proceeding six months, it has been amended to read six approaches, tracking a course and a hold in the preceding six months. This was done because in some parts of the country it is really hard to get six hours of flight solely by reference to the instruments. I guess the Fed's figured that if you can handle a hold, most of the rest of the procedural things would be easy.

Still, most savvy instrument rated pilots take an IPC once a year whether they plan to use the rating or not. At least you scrape some rust off of little used skills.

The problems associated with this are really food for thought for the instructor who gets the IPC assignment. Do you work the student back to a level of competency he has not seen since he received his (her) initial instrument training? Do you follow the letter of the regulations, fly a few approaches, a hold and a missed approach and call it a day?

I have many students who come back to me year after year because I wring them out and force them to fly to a higher standard than "just O.K." and I have a few students I know will never go actual IMC on purpose, and really want to just be sure they can keep the dirty side down should they accidentally find themselves inside a cloud.

There is little material published to offer guidance to the instructor to make that assessment.

One thing I have a real problem with (as an instructor) is the pilot who is unprepared for an IPC. That's right, unprepared. If you show up for an IPC and you haven't flown a lick of instruments in the last 12 months, much less looked at an approach plate, what are you telling that instructor? Are you saying, "This is a regulatory formality, so lets just get on with it, I'll do the best I can and it will have to be good enough?" Or are you saying " Take me out and teach me something!" Even the best instructors, who know their students can't read minds. If you are going to put the time into doing it, why not get something from it?

Here's a suggestion that will allow you to get a lot more out of your IPC (or BFR for that matter) and impress the hell out of yourself and your instructor.

Before you go for your ride, grab a buddy and fly a practice session. Put on the hood, let your buddy serve as safety pilot and go flying.

Bring a list of maneuvers you should expect to find on a typical IPC. And do it in the plane you plan to take your IPC in.

A typical practice program should look something like this;

Take off and depart the traffic pattern area. Then put on the hood. Have your safety pilot vector you the local practice area.

Climb to a safe altitude. Practice 45° left and right bank turns. Did you stay within the tolerance for the ratings you hold? Next, MCA, clean and dirty. Include 90° of heading change in the MCA configuration, in both directions.

Recover to cruise. Were you able to maintain altitude plus or minus what is in the PTS for the rating you hold? Then, move on to imminent stalls. Not full stalls, imminent stalls. That is to say, at the first sign of a stall, recover. What's that you say? That's not on the PTS for an instrument rating. You would be correct.

Now, move on to a hold. Any hold will do. An intersection hold on an airway, a DME fix, I don't care. Just go to the fix, enter the hold, time your turn, and adjust your outbound leg to give you a 1 minute inbound leg. Did you get the entry right? How many turns did it take for you to adjust your inbound leg length to 1 minute? Did you remember to adjust your speed to holding speed? Do you know your fuel burn in the hold? Did you remember to start the clock at the correct point?

Next, go find a non-precision approach. Preferably one that you don't know all the fixes and turns by heart. Down to minimums and the published missed please. How are you doing so far? Got an autopilot? Shoot a coupled ILS approach, followed by the same approach by hand. If you are a multi engine pilot, have your buddy "fail" an engine inside the marker by retarding the throttle on one engine. Do the drill. Simulate zero thrust. When was the last time you practiced a single engine ILS to minimums? When you get down to mins don't go around, flip up the foggles and land the plane. How ya doing now? Were you able to land using normal maneuvers? Don't quit on me. Try a zero, zero departure. Have you practiced one since you did your initial instrument training?

We're not done yet. If you have an IFR approved GPS in the aircraft, do you know how to use it? If not, there is no time like the present. That little box can save your life. Go shoot a GPS approach with a circle to land. If you're multi rated, do it on one engine. How'd ya do? Did you go below circling minimums?

Is all of this just too easy for you? Turn it up a notch. Shoot the same approach again but cover the artificial horizon.

Now ask yourself, " If I was an instructor, would I accept this performance?" Then ask yourself, " If you had your wife (or husband or significant other...) and kid in the plane and had to do this for real, would we all be here reading about your adventures or would you be a sad footnote at the end of an NTSB report."

If you answered "no" to any of these questions, if you busted minimums, if you can't hold heading or altitude within those minimums prescribed on the PTS (remember, those are minimum standards) then you need more work.

Don't get me wrong here. I am not suggesting that any of you need to fly to ATP standards, but if you hold a private pilot certificate with instrument privileges, being able to fly to private pilot minimum standards would be nice. If you are a commercial pilot, commercial pilot minimum standards should be your mark.

For most of us, it is not only O.K., but normal to be weak in some areas. Lets face it, none of us except those who fly professionally can get enough time in the air to be practiced at all aspects of instrument flight. And if you're reading this and thinking "He's got to be kidding, nobody does this" you would be wrong. Even those of us who do fly for a living go out and practice before a check ride, which we take every six months whether we need it or not, and everything I just described plus a few things I haven't are on the test.

Of course, there is nothing like actually doing it to sharpen your skills to a razor's edge. If you have the rating, why not consider using it once and a while. Don't want to go alone? There is no shortage of instructors willing to go with you at every flight school at every airport everywhere. You would be amazed how just one three hour flight every three of four months in the ATC system will sharpen your instrument flight skills. As an alternative, there are simulators at flight schools everywhere, where you can buy a few hours with a simulator instructor. The advantage there is you can practice things like failed gyros, severe wind shear, and other situations you can't safely do in flight.

When that is done, if you can answer yes to the questions we asked above, then you are ready to go take an Instrument Proficiency Check ride. If you can honestly answer yes then there is little an instructor or real life can throw at you that you won't be able to handle. When you are done, you will not only feel like you are Instrument Proficient, you will be Instrument Proficient, and not just Instrument Current.

This piece of wisdom had come to me long after I had become an instructor, and it was something my flight instructor never told me.

TMFINTM#109 -

The Biannual Flight Review

The Biannual Flight Review. Every two years whether you need it or not, the Federal Aviation Regulations require you to go out and find a Certified Flight Instructor and take a BFR.

The Feds developed this rule to insure that every pilot gets some remedial training in a reasonable time frame. Without it, you might find yourself sharing the traffic pattern with a pilot who last saw a flight instructor when Jimmy Carter was president.

Some pilots approach the BFR with all the disdain reserved for a tooth cleaning or filing your taxes. Others look at it as an opportunity to advance their aviation skill set, or at the least scrape the rust off of the one they already have. I am from the latter group.

For many years, I would simply work toward the next rating in the food chain, planning to complete it at the end of the two years. A new rating certainly meets the requirement of the BFR. That worked for a long time, because there are lots of ratings to get. But flight training for a rating is expensive, and money is always in short supply. One year I did a Seaplane Rating. What a blast. Possibly the most fun you can buy in aviation for under $700.00. A glider add-on rating costs about the same, also a lot of fun.

The Fed's don't give much guidance to us instructors on what to do for a BFR, but a cursory "once around the patch" is not what they had in mind.

I try to tailor my BFR's to the type of flying that my student does. If my student is flying into rural, uncontrolled airports on nice days, I'll take him to a Class B airport. If he or she is only operating off long paved runways in high traffic density areas, I'll take them to a grass strip to work on soft field procedures. If I don't particularly know them well I'll always ask if there is something they want to work on.

But every BFR I give includes a basic oral quiz on the aircraft they are flying. Since some of you are aircraft owners and fly the same plane all the time you all know all of the V speeds for your plane, right? If maneuvering speed changes with weight on your aircraft, you know those numbers too. Usable fuel vs total fuel? Zero fuel weight?

If you're flying a twin, can you tell me the accelerate stop distance at gross weight on a standard day? If you're flying a complex single or twin, do you know your gear speeds? Oh, and when was the last time you actually did a weight and balance problem? For that matter when was the last time you worked a flight plan?

When I do this, I'm not really trying to teach the student anything. Just pointing out that time has a negative effect on certain, unused skill sets.

Once we are out in the airplane, I like to watch rather than talk. Does the pilot use a checklist? Maybe you fly a simple airplane that you have owned for years, and you feel that you just don't need to use a checklist.

About 10 years ago I used to rent my J-3 Cub out to qualified pilots. One of them, a thirteen thousand hour airline captain came in one day for an hour of flight time in the Cub. About 10 minutes after I hand propped the engine for him (this Cub had no electrical system or starter) he came taxiing back in, obviously shaken. "It quit on me! I got 50 feet in the air, half way down the runway and it quit on me! There is something wrong with this airplane!"

There isn't much to go wrong on a Cub, and it was still running when he taxied back in. "Did you do a run up?" I asked. " Sure, ran up fine" he responded. I then look at the fuel shut off valve, which was 3/4 of the way to the closed position. It had allowed just enough fuel to flow to the carburetor for taxi and run up but when full power was applied for takeoff, the engine ran lean and choked. As soon as he retarded the throttle, it ran fine again. "It won't fly if the fuel is turned off" I said. He got very defensive. "Ah....., ah, that's really dangerous, that should be on a check list or something!" he stammered. It is. Along with the other three items, controls free and correct, mag check and carb heat, fuel to on is the first item on what passes for a check list in a Cub. Obviously, this pilot felt that the Cub was so simple that he could fly it without the use of a checklist, and it almost killed him.

So I like to see if a pilot uses a checklist. If not, my attitude about the flight changes. Once airborne, I like to do a standard air-work package that includes slow flight, stalls, and a simulated engine out. Again, I'm letting the pilot evaluate his or her own performance. If the aforementioned checklist was skipped, I might throw in a gear failure in a complex airplane, or an engine failure in a twin. If a pilot holds an instrument rating, even if they are not current, I might put them under the hood for some basic air work and a simulated gyro failure.

If the student is doing well, my work is nearly done. If not, I'll find an airport to land at, get out of the aircraft and grab something cold to drink. I'll ask the pilot what he or she thinks of their performance so far, and what they would like to do about it. Most elect to work a little longer and get it right.

If you are getting ready to take your BFR, go pull out a copy of the PTS (practical test standard) for the rating you hold. Go fly a flight and try some of the maneuvers you haven't done in awhile. Maneuvers such as turns on a point and S-turns across a road for example. Are you commercially rated? Go practice chandelles and lazy eight. Can you do them to PTS minimum standards? Been flying GPS direct-to everywhere?(who hasn't), go intercept a VOR radial and track it inbound and out. Instrument rated? Go track an NDB bearing.

Are you happy with what you see? Are you flying to minimum standards? O.K., go take the BFR. Not impressed? Practice a little. Remember, a BFR is just a review of your flying ability, not recurrent training. There is no way an instructor can uncover all of the flaws in your flight proficiency during a BRF. That's your job. Use the BFR as an excuse to go out and polish up your act. That little tip is something I learned through experience both as a student and an instructor, because it is something my instructor never told me.

Things My Flight Instructor
Never Told Me #110

Out of Control

A fter you read this story you will probably say to your-self...

"I've been there." Fly long enough and you will find your-self climbing into an airplane with a pilot you hardly know and he or she will do something that will just scare the hell out of you. You may find yourself wondering about both the safety of the flight and your personal safety all in the same thought. As an instructor I can tell you, I have most certainly been there.

A friend of mine came to me and told me the story of a trip with his boss. The boss was a non-instrument rated pilot. The weather was scattered at 4,000 and visibility in excess of 15 miles. My friend suggested to the boss that rather than bang around the cockpit below the condensation level, that a climb to 6,500 or higher would yield a better (read smoother) ride. The boss refused, opting to slug it out below the clouds. My friend went on to tell me that his boss had trouble holding heading and altitude, and appeared to be lost from time to time. The crosswind at the destination airport was distinct though not beyond the capability of the average pilot, but the boss expressed concern about his ability to land in it. My buddy was sweating bullets as the plane turned final.

The thought of "taking over the aircraft" was running through his head.

So what do you do, if you find yourself in a similar situation? This is something my flight instructor never told me. Here is my advice;

First of all, try not to blow things out of proportion. Keep things in perspective. If the weather presents a 4,000 ft. ceiling and 15 mile visibility, that doesn't sound very life threatening to me. In my friend's case, his boss's reluctance to operate "on top" is understandable. A lot of pilots are fearful of going "on top" because they are afraid of getting trapped, or worse.....getting lost.   Remaining in ground contact visually is natural for a non-instrument rated, low time pilot.

It would be fair to say, in my friend's case anyway, that at least during cruise, other than the turbulence, he was in no particular danger. That is, the turbulence wasn't dangerous per se, just uncomfortable.

As for your legal or regulatory position on the subject...... there is none except in a case where a pilot is physically unable to perform his/her duties as PIC. In a single pilot situation, there can be only one pilot. There is no provision to allow you to supersede your judgment over the judgment of another.

As for landing from the right seat; There are more than a few documented cases where non-pilots have safely landed small planes after the pilot became incapacitated, including an 88 year old woman who went flying with her 50 something son. He succumbed to CO poisoning and she landed the aircraft, with ATC help. So for anyone holding a private pilot's certificate, a right seat landing would likely be a non- event.

As for being there in the first place; I do fly for money, and I can tell you that I live and fly by the credo..... you can fire me but you can't kill me! I will not let money make a safety decision for me.   Commercial pilots are under that pressure all the time. Too bad, it comes with the job. But private pilots can just say no.

Once you accept, as in the situation my friend found himself in, you have to make the personal decision.

Is this situation really life threatening or am I just allowing my subconscious- control freak to roar in my ear? We all have it. You can't be a pilot if you don't possess the desire to be in control of a situation. THEN YOU MUST DECIDE; is the money going to make the decision for me? Yes or no.

Do you know this person? Is he or she reckless in life? Do you like their judgment on other subjects? Or is this a total stranger?

If you are willing to give up your job (as you should if your life is truly at stake) then demand that the pilot land the airplane right now, at the nearest airfield. Get a rent a car, or call your wife to come get you. While you are standing on the ramp, kissing your income stream good-bye as the airplane flies away, reflect on the situation. Be comfortable that you made a good decision and did not succumb to paranoia driven by lack of experience.

Do you feel the need to get that kind of experience? Becoming a flight instructor will certainly cure you of all of that. I have come within a fraction of a second of balling up a half-million dollar airplane and ruining a 20 plus year perfect safety record, while flying as an instructor. As an instructor, you have to let the students make the mistakes or they won't learn. Just how far you can let the mistake go, is a function of experience that comes from years and years of doing it.

My friend had a feeling that there might be a problem when he queried his boss about his BFR, which he had apparently just taken. He said it had taken a lot of time. That sounds like a red flag to ask more questions.

But for the record, there is precious little guidance given to instructors on that subject. In the BFR story in the previous chapter, I explained how I tailor the BFR to the type of flying that the individual does. Remember, it is a review. NOT A CHECK RIDE FOR A NEW RATING! As long as you perform within PRIVATE PILOT MINIMUMS, (or whatever rating you hold) you will pass. Instructors are all different. I train to a maximum performance standard, not a minimum published standard. Minimums are not good enough for me. But some guys will seek out low time or "easy" instructors, who, for what ever reason, lack of knowledge or the opportunity to make an easy buck, and will sign you off with minimum demonstrated competency. It is the pilot's fault as much as the instructor for looking for the easy way out, instead of admitting that he/she needs more work. Of course, more work usually means more money, and in the minds of some, good enough is ok.

My friend is a rated pilot. The boss knew his limits (to some degree) since they chose to do this trip on an absolutely CAVU day. I'm sure he figured he could handle CAVU.

In any event..... there is no precedent for taking over command, that is to supersede the PIC's judgment with your own. In the case of a CFI or an ATP..... the fact is that in the event of an incident or accident, it is usually the higher time/rated pilot that is at risk for certificate action, though again not always.

All you can do is be selective with whom you fly with and take the time to discuss this exact situation with your prospective pilot beforehand on future flights.

I used to have a buddy who owned a little Grumman. He was an excellent pilot. We flew together a lot. But he was a rotten passenger. The moment someone else was in control of the aircraft, he became very uneasy. If he ever sat in the back seat (only once with me) he would throw up, even on a perfectly nice cool smooth day. He would get all queasy and turn green. The moment he put his hands back on the controls, he was absolutely fine. Nothing was wrong. Obviously, he had some deep-seated control issues. I'm not a shrink, I'm a pilot, but you didn't need to have a PhD to figure that out.

Though I was a nervous passenger when I started to fly, 20 something years has fixed that. To me, the best compliment one of my fellow airmen can give me is to go to sleep in the plane while I'm flying. To me, that is like saying, "I trust you with my life." There are a few pilots I will close my eyes on.

In 20 plus years of flying and 14 years of teaching, I have only had to physically take control from a pilot once.

I was teaching Acro in my A-150 Aerobat. The student, a private pilot, was a professional fireman and built like the Incredible Hulk. It was a tight fit in that little Cessna. I was showing him how to push forward gently when inverted in a roll to keep the nose from falling through the horizon.

When it became his turn, he pushed yoke all the way to the instrument panel and kept it there. He also removed all aileron input. So there we hung, about 20 degrees nose up, inverted. "Release the back pressure and return the aileron to the stop", I said softly into the headset. Nothing. Frozen solid. A moment later the engine quit, because there are no inverted fuel systems on an Aerobat. I repeated myself, this time with a little more urgency, knowing we were going to stall. Still nothing. I pulled on the yoke, but he had it straight armed, locked full forward. Finally, in an attempt to get him to let go, I smacked him in the face with my aluminum kneeboard, which I was holding, and he released the controls. The nose fell, and we split S-ed out of the botched maneuver. I told him to return to the airport. Sweat was pouring off of him and he was angry that I had hit him. He could have easily ripped my throat out, but we went back to the airport in silence. I didn't charge him for the lesson, and I told him not to come back.

I had always wondered what would have happened if he didn't let go when I hit him in the face. I was wearing a parachute, and I suppose if you're not willing to use it you shouldn't teach acro in the first place. I probably would have bailed on him. I certainly wasn't going to let him take me into the ground with him. I guess we'll never know. The point is, that was a life-threatening situation. I had anticipated that it could happen, and came to terms with what I would have to do if it did. You need to do the same thing.

Things My Flight Instructor Never Told Me #111-

The Belt and Suspenders Approach

Here's an exercise. Go look in another pilot's flight bag and tell me what you see. The contents found in a pilot's flight bag tells a story. That story is the story of that pilot's experiences and offers an insight into what that particular aviator perceives is the most likely threat / problem / circumstance that he (or she) may have to deal with.

Six flashlights? Maybe that aviator had an electrical failure at night, or maybe he or she just fears one.

Spare sunglasses? Spare reading-glasses? Maybe the first pair had got left behind on the FBO counter.

My partner carries GE 33 and 331 lamps in his flight bag. Those are 12 and 24 volt bulbs commonly used in post lights and gear indicator lights.

I don't know what's in your flight bag, but two items you will find in mine is a hand held nav-comm, and a hand held GPS. I've carried them for years.

I had my first communications failure early on in my flight training. On one of my first cross-country flights, the single King KX -170 in the little trainer I owned at the time, ate the gears that change the frequencies. Unfortunately, it chose to do so right in the middle of the then Philadelphia TCA (now class B).

When I had left Morristown, New Jersey that afternoon, it worked fine. Half way to my destination on the Jersey shore,

I was proud of how I had negotiated my way through the busy Philly air space. Then, with no warning,... crunch, crunch, crunch.

Not sure if I should press on or turn around, I decided to return to my base at Morristown. I even remembered to squawk 7600, to signify communications failure to the radar controller.

Upon arriving at MMU, I circled the field just like my instructor told me, waiting for the light gun signals, which never came. I circled long enough to be able to tell what color shirts the guys in the tower were wearing before I decided to drop into the pattern and land before I ran out of fuel.

While I was tying down my plane, the follow me truck came roaring across the ramp. The line guy, a kid about my age jumped out of the truck and ran up to me, half out of breath. "The tower wants you to call them ..... right away!"

At this point, the adrenaline, which had been pumping with the boost pump on high since I crunched the radio, ran dry. I was tired. "Fine, what's the number?" I asked the frantic lineman.

This being the pre cell phone world, I finished putting the covers on my plane, and drove across the ramp to the FBO office where I could use the phone.

"Morristown Tower" the voice on the phone answered on the 1st ring. "Yeah, this is Grumman 9797U, you wanted to talk to me?" I said in my best, "I don't want to sound too annoyed" New York accent. "Yeah! You landed without permission! You can't land on my airport without permission! What the #$@% is wrong with you!" "Oh! Which one are you?" I said. "The controller in the red shirt or the one wearing blue?" I shot back. There was a pregnant pause on the phone. "What do you mean" he queried, his tone of voice changing dramatically. "I mean I circled the tower for more than 10 minutes watching you guys, waiting for a light gun signal like the AIM says I should. I couldn't wait any longer, so I landed". "Did anyone know you were coming?" he asked. "I don't know, I left here two hours ago, lost my radio in the Philly TCA, so I came back. I had 7600 in the transponder." "Hold on" he said. A million things were running through my mind, including how much I could live without this, when the controller came back on the line and said "O.K., never mind, have a nice day....." then click....he hung up.

The next day I ordered my first hand held radio. The early 1980's vintage ICOM was huge by today's standards, and it only comm-ed. The Ni-cad battery was expensive and unreliable. I carried it for nearly 10 years before I bought a newer model that ran on dry cells and added a nav function. The ICOM A-2 also came with a jack that lets me plug in my headset directly to the radio. In 20 years, I have had to pull it out of the bag exactly twice. The first time, just a few months after I had purchased it. On my first night cross-country, the radios in the rented Cessna just died. My instructor was very impressed when I reached down onto my flight bag and pulled out a working radio. The second time, again at night when I had a regulator run away and we were forced to shut down the electrical system or risk a battery melt down.

The other thing I carry is a portable GPS. In 20 years I have owned several airplanes that didn't have an electrical system of any kind so I got some use out of that investment. The first GPS I purchased was a Garmin 65. It was the size of a brick, but it ran on dry cells and like most handheld Garmins, it featured an antenna that could be remotely mounted. I traded up to a Garmin 89 a few years later, and a few years after that, I traded up again to a Garmin 92. The 92 is the same size as the 89 but uses a more powerful satellite engine, tracks more satellites, locks on faster and does a better job on battery consumption.

It has resided in the bottom of my flight bag for the last three years. The last time I had it out was to do a data update back in 2000.

In the last four or five years I have been flying some pretty decent equipment, much of it turboprop, and everything always works. That is not always the case for those of us who operate IFR in singles and light twins.

So here I am, 140 nautical miles off of Jacksonville, Florida on Atlantic Route 7 going to New York, when the always reliable Garmin 155 in the panel of the King Air I was flying, decides to have a little electronic temper tantrum. I have 8 passengers in the back eating lunch and talking about the Springsteen concert I was taking them to see.

I reached down into the bottom of the flight bag and pulled out the trusty Garmin 92. It hadn't been run for three years. Would it lock on? Would it find any satellites? I let it run for 20 minutes but it couldn't find what continent it was on. I went to the manual set up mode and told it to look in the South East United States. It started looking for satellites again. Meanwhile, I dialed in the Carolina Beach NDB and then the Dixon NDB, the next fix on my clearance. I silently chastised myself for not taking that GPS out of the bag and letting it lock on to the satellites every once in a while.

Twenty minutes later, the little 92 locked on and gave me a position report. It verified that we were within 2 degrees of where we were supposed to be. Situation averted, and no one in the back knew the difference. We flew on, picking our way through a line of thunderstorms, shot a 400 and stinko approach in the pouring rain onto TEB, using the little Garmin for secondary position information.

A few hours after we landed, the skies cleared, and Springsteen was awesome. A few weeks before his 54th birthday, Bruce rocked with 40,000 of his closest friends in Giants Stadium, just like he did when I used to go listen to him as a teenager, at a Jersey shore bar.

The next day was glorious, but of course, we weren't flying that day. Nope. Departure was set for Monday 7a.m. Just as when we arrived, the weather sucked and we flew the D.P. for real, not breaking out of the clag until well south of Philadelphia on the way to 28,000 feet. The little handheld worked like a champ, locking on in under a minute. We flew all the way home, 4 1/2 hours, without missing a beat or killing the batteries.

Was it a life-threatening situation? No. Did we have to have it to complete the flight? Again, no. The plane had dual everything else. Was it nice to have the GPS back up flying into and out of the busiest airspace on the East Coast? Absolutely!

The same could be said about the comm. The failure I had experienced happened years ago. Had I had a handheld on that day, it would have been a routine, no stress situation.

When I got home from my New York trip, I decided that the next thing I wanted to do was update the database in the Garmin 92. The last time I did that, you needed to send the unit back to the factory. Now it is all done on the internet.

All you need is a serial cable to attach the GPS to your computer and access to the internet. $35 and 15 minutes later I had a current database installed in my GPS. Ya just gotta love the technology.

So, back into the flight case, along with a 12 pack of fresh batteries for the GPS and the ICOM A-2 waiting for the next time, if there is a next time, that I have an in flight nav or comm. failure. No, I'm no boy scout, and yes, my flight bag is heavy enough, but the ability to communicate and navigate in the event of a total or partial failure mode lets me sleep at night.

While I have heard the cell phone argument, and agree it might work, I'll ask if you have you ever tried to hear a cell phone in a running piston airplane, much less fly at the same time?

If you look at the accident reports, there is no doubt that a handheld radio or GPS would have changed the outcome of more then a few instances.

When I learned to fly, this type of portable technology just simply didn't exist. In 1980, what passed for a handheld radio was made by Terra, and it was the size and weight of a one-quart milk carton.

But now, I make a point of showing my instrument students how a few hundred dollars worth of technology can take a potentially catastrophic situation and render it totally manageable.

Maybe carrying all that equipment seems like overkill. Particularly, since it is almost never utilized. But that's the point isn't it? Like the guy who wears a belt and suspenders, (that would be me) I feel that the redundant capability, even if unused, reduces the chances of finding myself in a potentially fatal situation.

The Organized Pilot

It is not uncommon for a new pilot to approach aviation just like his or her instructor does, since the instructor is the primary source of influence on the newly minted pilots thought processes. Over time, each pilot develops his or her own style of doing things and develops his or her own sense of cockpit organization and preflight preparation.

For me, a major source of in flight stress, especially when engaged in single pilot IFR operations, comes from lack of preparation for the contingency that now confronts me. You know, the kind of stress you feel when you get a full route clearance in flight that reroutes you to waypoints you never heard of, in a part of the world you've never been to. You diligently copy the clearance, and then set off on the task of finding the route or intersection on a chart. But which chart? Low altitude, high altitude, approach plate, STAR? The paper shuffle in the cockpit can get furious.

I have always believed in serious cockpit organization. In the last few years, I have been flying for money, and have incorporated a few new techniques into my routine, which have greatly reduced the overall in flight stress level, while allowing me to be a more efficient pilot on a number of levels. This is how my flight prep and in flight routines have evolved.

Going Electronic; A major asset to my preflight planning, weather gathering, weight and balance computations, and flight log preparation comes from my computer. I bought a little Sony Vaio 505 laptop. This little beauty is so small it easily fits in my flight bag. I loaded it with RMS Flightsoft™ flight planning software. If you haven't tried one of the electronic flight planning programs, you just don't know what you are missing. This type of software elevates flight planning to a new level. You put in the departure and destination airports and it will automatically give you the preferred route if there is one, depending on your altitude. It will go out and get the weather, current and forecast, and will overlay it on a map of the route. It will allow you to play "what if" with the wind, identifying which altitude will give you max range vs shortest time. Doing a weight and balance is a snap, and it will even tell you how to fix an out of balance situation. It will generate current VFR charts with your route clearly depicted in a number of selectable scales and will depict the most current TFR areas.

If you fly several different aircraft, it is easy to set up profiles for each one. It will even track operating costs and pilot flight times if you so choose to use those features.

I like the flight log, which you can customize to fit your personal taste. I generate a flight log, and a waypoint sheet, which lists all the waypoints, in order, with coordinates and frequencies for each flight. I also generate a flight plan form from the same program. You can file that form on line right from the laptop, eliminating the need to talk to a briefer, though I rarely use that option.

I use the flight plan form, which is generated at the top of an 81/2" by 11" sheet as the top sheet to identify the leg I'm flying. I attach the flight log and waypoint log under it.

Then I generate all the approach plates, DP's, STARS, and airport diagrams for the departure airport, the destination airport and my alternate using Jeppview™. I know, for 20 years I flew with NOS plates. Actually, if you look at the Jepp products and the NOS plates, they are starting to look pretty similar. Jepessen has always been very expensive, and I hated doing the updates. I like the convenience of NOS plates, which get discarded every 56 days and I get a new book.

But Jeppview eliminates the need for updates. You get a new disk every 56 days, you load it in the computer and presto, all of your plates are current. Yes, it is still expensive, but it is less so then paper plates and much more convenient. I still carry a printed approach plate book for Florida, which is my home area, but I rarely use it.

One nice thing about electronic plates some of you may appreciate is the ability to output them in larger than standard format. That is, you can output a plate in 8 1/2" X 11" size, making it a whole lot easier to read, especially at night. The program gives you several output options. I usually put two plates on a page, just to save paper.

All of this gets attached to the back of previously outputted flight plan form, flight log and waypoint sheet. The last thing I make is an enlarged copy of the destination page from my AirGuide™ Flight Guide.

Like many of you, I have carried that little brown book with me for years. I like the airport diagrams, and the on field services information. Yes, Jepps supplies it as well but it is not in diagram form and the Airguide format is very user friendly. I usually enlarge it to fit on a single page just to make it easier to read.

This entire package goes on a lightweight aluminum clipboard. I use the bottom 2/3rd's of the flight plan form page to copy the ATIS and subsequent clearance.

For years, I used to highlight my route on my enroute charts, both VFR and IFR. I carried a three-color highlighter, for use in the event of a reroute in flight. In the last few years, I have been flying with my partner, who became incredulous when I asked if I could write (or more accurately, highlight) on his Jepessen enroute charts.   I have since taken to using highlighter tape, which works even better because you can peel it off and reposition it in the event of a reroute.

Since 90% of what I have been doing lately has been IFR, I invested in an Air Chart Systems VFR chart Atlas, to keep me legal in the event I choose to cancel IFR early, or reposition to a nearby airport VFR instead of IFR. For $80, you get every VFR chart (WAC) in the country plus a number of terminal charts and updates for a year. I probably don't need it since my flight planning program generates current VFR charts, but when operating under Part 135 you need to have the source of navigation information approved by your principal operating inspector at the F.A.A., and ours is not that impressed by computers. For part 91 operations there is no issue. In fact, for part 91, there is no requirement to have a printed approach plate in the airplane, only a text description of the procedure, which means you could pull the plate up on your laptop and use it right off of the screen.

In the cockpit, the charts are arranged in order of use in a chart wallet. I don't even have to look at it to get the next one in sequence out. If I have time in flight, I will put the unused chart away in the proper place, if not, it goes into a side pocket in order of use.

Does all this sound ultra anal? Overkill? Until I started doing it like this I would have agreed with you. But since then, I have noticed a marked decrease in the amount of cockpit scramble that occurs when ATC decides to see just what kind of airman you really are.

A few years ago we were flying in and out of Philadelphia International and we got 3 full route clearances between the ramp and the end of the runway. Two more came after departure. I was a new copilot on a Citation that day, and so far behind the clearance delivery I was embarrassed. My captain just smiled. He had flown out of this airport more than a thousand times and knew the departure routes by heart. I on the other hand scrambled to find the secret waypoint and decode the clearance.

A few weeks ago I was there again, this time as captain, and this time they couldn't shake me.

Last weekend I did three trips to Freeport, Bahamas on the same day. Each time I departed for the States, they gave me a different departure route. Out there, there is no radar (most of the time), so it is strictly pilot nav. Despite the heavy accent and my personal unfamiliarity with the local procedures, they couldn't lose me. My buddy in the cockpit that day, an experienced seaplane pilot was in the right seat. He is just the opposite of me. He goes everywhere VFR at 1,000 feet. I had given him a multiengine rating some ten years ago. When the flight was over he commented at how calm and relaxed I was in the aircraft as opposed to my demeanor ten years ago as his instructor. At first I didn't think anything of the comment, but then I realized that I had lowered the level of in-cockpit stress to the point that it had become obvious to another pilot who knew me well. That alone is worth the all the effort.

The bottom line? Develop a preflight planning and cockpit routine that works for you. There is no shortage of organizational aids and tools available to you. Avail yourself of the electronic technologies because they make a tremendous difference.

Constantly keep refining your act to incorporate new skills and tools as they become available. The net result is that you will be more relaxed and confident in the cockpit and you will enjoy your flight time that much more.

A View From The Right Seat

Most of the time, when you hear me talk about a view from the right seat, I am speaking as a flight instructor. In this story, I am not speaking as a flight instructor but about the right seat occupant, a passenger.

For most of us who own an airplane, that passenger is usually the significant other in your life. In conversations with other pilots, I have heard this individual referred to as the "non flying spouse".

I have been an aircraft owner for more than 20 years now, and my wife has flown everywhere with me.

When we were young(er) and had no kids, her attitude about flying was radically different than in the post-children reality that is my current existence.

My wife grew up around airplanes. Her dad was senior flight test engineer for Grumman Aerospace during the hey-day of post WWII military aircraft development. When she went to work with dad, she got to play with the F-14's, A-6's, and the occasional Lunar Module. As a kid she was "forced" to go to the Grumman company picnic to see the Blue Angels perform, every single year. But while she may be in love with me, she has never been in love with aviation.

If I were to say to her, "hey honey lets fly out to the Hamptons for the weekend," she would race me to the plane. But if I were to say, "hey it's been a tough week and it's a beautiful night, let's go for a ride in the plane," she would find 49 other things to do that were far more appealing to her.

Mind you, she would never stop me from going, either alone or with a flying buddy, just not with her.

Over time, she made an effort to gain enough flight experience that she could handle an airplane fairly well. As we progressed to larger aircraft, she demonstrated a natural ability to fly them, if not well, then well enough in an emergency.

Her primary concern is probably typical of most non-aviators forced into the "non flying spouse" role, and that is, if "he" has a heart attack, "I" don't want to die too.

Despite my lifelong assurances that I was never going to die in a small airplane, if for no other reason than there was no way that I was going to let her next husband get my money that easily, my wife remained, and still is, an uneasy passenger.

The arrival of our son has only added to her uneasiness. When the three of us are in a plane together the tension is so thick you can cut it with a knife. My kid picks up on my wife's tension and he stresses out. It is no fun for me at all.

A few years ago we went to Tampa for a day at a theme park. It was severe clear when we departed West Palm Beach, less than an hour away. Upon arrival, a marine layer had moved in and I was forced to file a pop-up, local IFR flight plan. Our destination airport was a VFR only airport and at minimum vectoring altitude, we were solid in the clouds. The controller asked my intentions and I asked for the ILS to Brooksville, about 20 miles away. I asked my wife to look up the approach plate, a task she had not been briefed on, and she couldn't find it. All the while, she was asking me why we are doing this and going there, and what about the rent a car etc, etc. My kid picked up on the cockpit stress level and came unglued. When my kid gets a little older and can handle the plane himself, that should all go away. In the meantime, I take it one flight at a time working toward easing that anxiety.

In the last few years, VFR has been no problem; only IFR and crummy weather operations set them off. For those trips, the solution to the anxiety has come in the form of a second, qualified pilot on board.

I find no particular consolation in the knowledge that it is not just my wife and my kid that do this.

In the 14 years I have been a flight instructor, I have given dozens of "Pinch Hitter" courses to the wives and kids of my students and buddies who were going through the same thing. By empowering the non-flying spouse with the knowledge and more importantly the confidence to land the plane in an emergency, it has made flying together not just tolerable but enjoyable.

I had one buddy, a hangar tenant of mine, who had not one but two airplane accidents. Both times his wife and one of his kids or the other were in the plane. The second incident was a bird strike. They hit a baby eagle on final approach to our home airport. His wife was convinced that had it come through the windshield and taken her husband out, she and her kid would have perished as well.

This college educated professional couple had defied the odds and had not one but two aviation accidents in the about 500 flight hours that this man showed in his logbook. Logic and the law of averages argument just didn't work for her anymore.

She came to me with one purpose; teach her to land the plane.

Two weeks and 10 flight hours later, she could land that thing as well as anyone. In the middle, a few hours of ground instruction reinforced what we had done in the plane and reviewed communications procedures.

The fact is that she was done in five hours, but she wanted the extra dual to be absolutely sure she would and could perform. What I didn't know was that she had neglected to tell her husband she was doing this. They went out flying for dinner one night and he was absolutely shocked when she asked if she could "try a landing". He agreed, she greased it on, then, fessed up to the dual instruction.

When I asked her why the secret, she told me that she thought that her husband would be insulted that she didn't trust him with her fate. Whatever. That scenario never crossed my mind.

Another buddy asked me to teach his wife simply because he didn't have the self-control to keep his hands off the controls and let her fly. That, I could totally understand.

The most interesting case for me was the number one captain at Eastern Airlines. He was retiring and was dead set on buying a Cessna 195. The airplane he ultimately purchased had a turbocharged Jacobs radial engine in it. He planned to winter in Florida and spend the summers in Washington state.

His second wife was not an aviator. What's more, he insisted that we instruct her in a tailwheel airplane since that was what they were going to fly.

For those unfamiliar with the Cessna 190-195, they are classic old radial-engined monoplanes. They are virtually blind on landing and have a nasty reputation for being difficult to handle on the ground.

I spent 35 hours in a Cub with this woman and she still hadn't soloed. To be sure the problem wasn't me, I sent her out with another tailwheel instructor for another 8 hours, and still, not even close to solo. I'm entirely convinced that had I been allowed to instruct her in a nose wheel equipped aircraft, she would have done it in 10 hours, but her husband wasn't hearing any of that.

They flew off to Washington State and I never heard from either of them again. What does that go to prove? Only that the non flying spouse needs to have a motivating factor to want to learn to do this, and not simply to do it because the flying spouse wants them to. Whether that motivation is self-preservation or self- satisfaction (if "he" can do it, "I" can do it) it really doesn't matter.

Though they love us and want to share our time and adventures, it is most likely they will never share our passion for aviation. The flip side to this is that if the non-flying spouse develops a new level of comfort in the airplane, you are going to get to do more flying.

Pinch Hitter courses have been around for a long time. Personally, it has been my experience that a trusted friend or instructor who has done this sort of thing before, is more likely to bridge the confidence gap then some young flight school instructor. I'm not knocking them, but it is hard to put the trust of your life in someone young enough to be your kid, and take it seriously.

All of the flying should be from the right seat, since that's where it would have to happen for real. And it should be done in your plane, not some flight school plane, even if it is the same make and model. Knowing where all the switches are, and how the radios work, are part of the deal and no two planes are identical, especially older aircraft that have undergone extensive equipment retrofits.

When I teach a non-flying spouse, I keep it super simple. If the plane will land with the flaps up, let's leave them there. I believe the less re-configuring of the airplane that has to be learned, the better. I'll go find a nice big runway and "vector" them to it just as a controller would in a real emergency. I teach them how to answer the questions that the controllers will ask, and get the information they will need from the radios and instruments on the panel. But mostly what I do is teach them that the plane will fly just fine by itself. If you have an autopilot, I teach them to use it. The key here, as in any life-threatening emergency is to relax and allow the self-preservation mode each of us possess to kick in.

As a closing thought, consider this; most of us wouldn't think twice about spending money to increase the usefulness or utility of our aircraft. Money spent on addressing the fears of the "non-flying spouse" will do both.

Bad Service

Most of the time, the quality of the service you get from air traffic controllers is consistent and good. But, during a period of time spanning eight or nine months I had received what I would have to call just plain old " bad service" from a number of air traffic controllers. I'm not talking about the errant call sign, or the "sorry, unable" you might get when asking for radar service while flying VFR and they are busy. No, I'm talking about bad service while on an IFR flight plan in IMC for real.

What makes these instances so relevant, is in 24 years of flying and nearly 16 years of instrument flying, I can think of maybe two or three instances where I felt the controller legitimately dropped the ball. But three times in a year makes me wonder out loud "What's up?"

Fortunately, I had other pilots on board in all three instances, so I was able to get another airman's "take" on the events.

The first incident occurred on the way back from Atlantic City, New Jersey, en-route to West Palm Beach County Park Airport, locally called Lantana (LNA).

The first leg from ACY to CRE, Grand Strand, North Myrtle Beach was wonderful. A late spring cold front had blown through the day before leaving clear cold skies, unlimited visibilities and yes, a tailwind.

At CRE we refueled, check the weather and filed our final leg home. I typically operate that airplane, a normally aspirated, J model Mooney, at between 8 and 11,000 feet. I filed Victor 3 and 10,000 feet. The local procedure when arriving for

Lantana from the north (and being RNAV or GPS equipped) is typically the airway to Vero Beach then direct Stoop intersection, intercept the PBI 359 radial and track it inbound. You're then vectored to the final approach course, the 002 radial. You can expect to cross Stoop at or below 8,000 feet. The crossing altitude on the approach at the FAF is 2,000 feet.

Approaching Vero we were cleared direct Stoop, told to cross at 8,000 and to report leaving 10,000 feet. Stoop was some 32 miles down the road, so I was in no immediate hurry to descend since I only had 2,000 feet to lose. At this point I was talking to Miami Center.

The weather was still o.k., with some low coastal scuddlies blowing in on the north east wind and some small convection were mixed in, mostly over the warm water, which by the way is where Stoop is. 15 miles from Stoop I reported leaving 10,000 for 8 and was told to contact Palm Beach Approach. So far all was normal.

Directly ahead, parked right over Stoop, was a small convective cell not more than a mile wide. But the top was well above my altitude and building slowly. A few sparks were coming out of the bottom, which my storm scope confirmed. The rain shaft was solid. I simply could not see any good reason to fly through it. Being IFR - GPS equipped, I can see my cross track deviation to the 100th of a mile. I decided to pass the shower on the left. That put me just a hair over 1 mile off the centerline of the airway.

Three miles from Stoop I was passing through 8,300 and my instrument student in the right seat was asking if it was o.k. to get down before passing the fix or did the controllers expect you to arrive at the intersection and the altitude at the same time, when the approach controller called. "Confirm you are direct Stoop intersection" she asked, which I did." "I show you right of course," she said. I said "Yes, by maybe a mile or so." At that point she exploded. "You can't deviate without asking me! Now I need you at 6,000 now, report reaching!" Assuming she had some outbound traffic climbing out in front of me, I popped the speed brakes and down we went at 1,500 feet per minute. Less than a minute later she asked my indicated altitude which I reported at 6,800 feet, and just as I called to report 6,000 she handed me off to the next sector controller who issued the direct PBI, expect vectors to the visual clearance.

The ATIS was reporting 1,600 broken and raining at PBI just 5 miles north of Lantana, so I was curious about the visual approach clearance. Normally, they fly you right over the top of PBI and you start your descent into Lantana, just as if you were flying the VOR/GPS 15 approach.

We got vectored west of the Palm Beach International Airport, and descended to 2,000 feet. This made me uncomfortable because of an antenna farm with tops at 1549 agl. Once south and west, they vectored us direct to LNA. The problem with that was that we were IMC in rain and popping in and out with just glimpses of the ground.

The approach controller then cleared us down to 1,600 feet, the minimum vectoring altitude over the airport. At this point I was getting pissed, since it appeared we would be asking for vectors back from where we came so we could shoot the approach. I should have insisted on the approach to start with. Just as I keyed the mic to ask, we popped out into a hole, and I could see the runways below me. I cancelled IFR, popped the speed brakes and dropped down to land. The bases were 1,100 feet. The weather was moving fast and by the time we taxied in it was raining hard on the airport.

My student had more questions than I had answers.

Why did the approach controller give you a hard time about being a mile off centerline? Why the slam-dunk at Stoop from 8 to 6,000? And why did they issue the vectors for a visual approach when the weather was clearly below MVA?

All of them were good questions. So, to satisfy my own curiosity as well as my students, I called the Tracon supervisor on the phone. Got the number from that little cellular pilot phone book. Great tool.

I first asked if I didn't own the airspace 4 miles either side of the centerline of the airway. He said that yes I did. I asked about the slam-dunk from 8 to 6,000 at Stoop as well. He just didn't know why that instruction was issued. Then I asked why we were given vectors for the visual when the weather was so low. I thought that the ceiling had to be 500 feet above MVA in order for them to issue a visual. He answer was yes and no. Because LNA had no weather reporting on the airport, if the controllers had reason to believe that the weather was above the minimum to issue a visual they could. So I asked what led them to believe that was the case. He told me that there were numerous targets in the pattern shooting touch and goes so it must have been VFR. I wasn't liking that answer too much and I asked a few more questions. Then he revealed that PBI was landing on runway 32 and if they issued me the approach into LNA from the north, they might have to stop operations on runway 32 at PBI until I cleared the approach. I asked him how that was fair and he said it wasn't, "you just got some bad service".

Two days later, a flying buddy of mine going into North County Airport (F-45) got the same deal. The GPS approach from the east tracks across the missed approach path for runway 32 at Palm Beach, so he was issued radar vectors for the visual, only to end up missing and turning around to shoot the ILS to runway 8. When he told me that story, my experience of two days earlier was put in perspective.

Fast-forward a few months. I'm now working with an IFR student also in a Mooney, on an IFR cross-country to Ft. Meyers, Page Field. We are on an IFR flight plan in solid IMC. Just west of La Belle VOR (LBV) we are switched to Ft. Myers approach. That controller seemed surprised to be hearing from us. A few minutes go by and he tells us he's got nothing on us and what did we want to do. I explained this was a training flight and asked for "The ILS to runway 5, the VOR to runway 13 and the NDB to runway 5 and then we will go away" I promised. He cleared us for radar vectors for the ILS 5. While setting up the radios I noticed the compass locator-outer marker wasn't indicating. I wasn't sure if it was the radio or the beacon but I wanted to see if my student caught it so I said nothing. We were in actual IMC conditions and he was struggling with his situational awareness. He noticed the LOM not pointing and asked me about it. I told him to ask the controller. "It indicates it is working fine up here" the tower controller said, leaving me thinking our ADF died. Another aircraft on the freq. heard the conversation and chimed in confirming he wasn't receiving it either. After a brief "standby" from the tower controller, the controller came back on the frequency and confirmed that in fact it was the beacon that was in-op. That made me feel better about our NDB, but left me wondering what else the people on the ground don't know about. My student dialed the marker in on the GPS and down the pipe we went. On the missed approach we decided that since NDB approach was out, we would shoot the VOR 13 and head home (east). With the clearance for the VOR 13 approach the controller gave us "runway heading to 2,000, left to 050 and contact departure on the missed".

When we checked in with departure the controller bit our heads off. "Weren't you assigned runway heading?" he queried. "A-firm" I replied. "Then why are you heading 130?" He shot back. "Cause that is the runway heading" I replied. Silence filled the frequency. Obviously, he thought we were shooting the ILS 5 or the NBD 5, not the approach to 13. He was pissed, but not at us. "You're correct, turn left to 050, climb and maintain 5,000."

I spent the rest of the ride home trying to convince that student that it doesn't always happen like that but he wasn't buying it. We flew several more IFR X/C trips together before he would even entertain the notion of going for the check ride, even though he was more than ready. Just some more "bad service?"

The last episode occurred a few weekends ago.

On a clear winter morning, we departed Palm Beach, Florida, for a VFR cross-country trip to X39, Tampa North. The forecast was for VFR. Indeed, the weather was CAVU on departure. Our route was basically direct Labelle direct X39. At 60 miles out I called Tampa approach and received a transponder code. Ahead we could see the edge of a cloud deck, so I decided to start a descent to put us below it. The ASOS at X39 was in-op and I could not raise St. Pete Flight Watch, but with 40 miles or so to go and the forecast for VFR I continued on.

Just south of Plant City airport (PCM) I could see the ceiling started to drop ahead. I called Tampa Approach to ask if I could have a local IFR clearance so I did not have to drop down to 1,000 feet in order to stay VFR. (Being unfamiliar with the location of the antenna farms in this area, I felt that VFR in 4 or 5 mile visibility at 1000 feet was not smart). He told me he didn't know where I was and that he had given me that code "a half hour ago". I pointed out it had been 16 minutes. He told me to contact approach on another frequency, which I did. When I checked in, I told him that flight conditions ahead looked IFR. I asked if could I get a local IFR clearance at MVA to X39, and if unable to land (descend from MVA to the airport and land in VFR), we would then like vectors to the ILS 9 at Brooksville (BKV) 13 miles north.

He said, "OK" then gave me a new squawk code and the instructions "proceed direct to X39 GPS direct or whatever you have, descent and maintain 1,600". I repeated the clearance, and told him we were IFR current and capable since he hadn't asked. He then told me to report flight conditions when over X39. When we arrived over X39 at 1,600 we were IMC. I reported that fact and asked for vectors to the ILS 9 at BKV since there is no approach at X39. To this point, we had not entered Class B airspace. We had been operating in the "3,000 ft" area to the east, and had been below it. His instructions next were climb to 3,000 feet, turn left to 280, expect vectors to the ILS 9 at BKV. As we climbed through 2,100 feet we broke out into the clear, and the remainder of the flight was uneventful. We broke out on the ILS at just under 1,000 feet, maybe 950, and I was able to cancel on the frequency as per the controller's request.

What I never heard, was "cleared to the Class B airspace". I also never heard a complete IFR clearance of any kind.

Normally, I file IFR everywhere. But sometimes, when going to uncontrolled fields with no published IAP on a day reported to be VFR, I just go VFR. The encountered weather was not forecast. At BKV there were several planes in the pattern shooting touch and goes, as they were at X39 when I had flown over it (as reported by the people waiting to pick me up). An hour and a half after I landed the weather was as CAVU in Tampa as it was in Palm Beach when I departed.

As an instrument flight instructor I am very familiar with the procedures at these local facilities. I had experienced what could be called "bad service" on a number of occasions in the last year or so, while flying IFR. But in this case, I suspect the initial controller just didn't want to deal with a "pop up" clearance so he put me off on the next controller, who handled it beautifully. I don't make a habit of "pop-up" clearance requests and am annoyed at myself for not asking for it sooner. But here at the Palm Beach Class C airspace, we do it all the time for training purposes. In fact, the Tracon supervisor at PBI told me when I phoned him to inquire as to the best procedure to obtain an actual clearance for actual IFR in order to shoot approaches with a student at PBI, he told me that they prefer it. (The pop up / local clearance).

If the initial controller at Tampa said "no", I would have landed short and filed or circled south of the class B, called flight service on the radio and filed a flight plan that way. But I have never been refused a local IFR clearance before so in my mind, those options were probably not going to be required.

I might assume that the second controller assumed that I had been cleared since I had a transponder code already so he didn't feel the need to re-issue the "cleared for the Class B" clearance. Though at the time it did cross my mind to confirm clearance into the Class B from the second controller, the frequency was busy. There was another aircraft, a twin, with a double voltage regulator failure in IMC negotiating with the same controller and I felt his situation probably took precedence over mine in the mind of the controller so I left it alone.

There was certainly no tension on the controller's part, no irritation in his voice, no admonishment to contact anyone upon landing, nothing like that. For all I know, the only person who noticed that I never heard the clearance was me. It is possible it was issued and blocked. The frequency was very busy and a lot of transmissions were being stepped on.

As a flight instructor, I drill into the student's mind that you MUST hear the words "Cleared to enter Class B," from the controller. I have, on more than one previous occasion, had to ask a controller after initial contact and being issued a squawk, if I was cleared for the Class B when I did not hear one issued.

In this case I suppose it is bothering me that I did not hear that clearance. I also feel that the initial controller telling me he didn't know where I was, when he had issued me a squawk code just 16 minutes earlier, is a scary thought. Why else would anyone call in and get a transponder code except for the controller to know where he was? The second controller knew exactly where I was and did a fine job of handling the situation.

The bottom line is this; Sometimes you're just going to get "bad service" and you need to be prepared to deal with it. If you fly the same arrivals and departures all the time you are familiar with the drill. If you are getting something else, be extra cautious.

If I had listened to myself and been a little more assertive, I would have insisted on the approach in the first instance and just not accepted the visual approach. I was so rattled by the controller admonishing me to ask for a one-mile deviation that I became complacent when I was issued the visual approach.

There was little I could do with Ft. Myers approach on that day. We followed the instructions; it just wasn't what they wanted.

As for Tampa approach, I probably should have asked for the local IFR on the initial call up. That would have given everyone more time.

The point is, incidents, which could result in a violation or an accident, don't always originate in the cockpit. Sometimes it is initiated by "bad service".

SOME THOUGHTS ON DENSITY ALTITUDE

Well here comes summer once again. If you are based east of the Mississippi, that means high temperatures and low flight visibilities, thunderstorms and high density altitudes.

Here in sunny South Florida flight vis doesn't get too bad, but temperatures on the runway can exceed 140 degrees F. The effects on aircraft performance in high temperature situations are the result of density altitude.

Some aircraft are actually prohibited from operating in these high temperatures. Some of those limitations are based on the fact that there is no performance data for pilots to plan from. The Citation I am type rated in is prohibited from takeoff in temperatures above +39°C ISA.

By definition, density altitude is the measure of air density. Do not confuse it with pressure altitude, true altitude or absolute altitude. It is not to be used as a height reference, but as a determining criteria for planning the performance of an aircraft. In simple terms, it is the altitude the aircraft "thinks" it is at. Molecules of warm air are further apart then in cold air. Further more, in high humidity situations, since two molecules can't occupy the same space at the same time, water molecules displace air molecules. So why would I care while flying here in South Florida where the highest surface elevation is the on ramp to the freeway?

Even on a day where the barometric pressure is 29.92 inches of mercury, when the surface temperature reaches 140°F, which it can easily do in the heat of the day, the aircraft "thinks" it's at 5,000 feet.

Look at the performance section of your aircraft flight manual and see what it says about temperatures above standard and how they affect aircraft performance. For time, fuel and distance to climb, typically you need to add 10% for each 10 degrees C above standard. If a standard day is 15° C, and 140°F is about 60°C, you need to be adding about 45% to your time, fuel and distance to climb figures. This becomes especially important when flying departure procedure with non-standard departure minimums.

A perfect example of what I am talking about is the TATES TWO and the WILTT TWO departure from Altoona, PA. The field elevation is about 1,504 feet. It is about 11.5 miles from the field to Tates intersection, which you must cross at 5,000 ft. That means you must climb 3,500 feet in 11.5 miles or 318 feet per nautical mile. To achieve that in a typical high performance single you would need to see a climb rate in excess of 525 feet per minute in a 90 kt climb. Can your aircraft do that when it is "thinks" it is at 6,800 feet already?

Remember that the performance figures in your book are for a factory new aircraft, flown by a factory test pilot in what can only be described as "factory air". Real world experience says add 10 to 15 percent simply because your engine and airframe are not factory new and you are not a factory test pilot. So you don't need to be flying in Colorado to consider the effects of density altitude on your flight planning. Not all obstructions are manufactured by nature either. Here in Miami, the Miami 6 departure features non-standard departure minimums so you clear the condominiums on the beach, east of the airport. They put the obstruction height at 1,300 feet, (not because the building is so tall but for noise) then you add 50 feet to that. You need to cross at 1,350 ft. Can you do that on a really hot day in your aircraft?

Density altitude has a minimized effect on turbocharged

aircraft because, unlike their normally aspirated cousins, these engines continue to make full rated power at high-density altitudes. But the wing still thinks it is operating at the high-density altitude.

One trick jet operators use when it is hot is to reduce the weight of the aircraft. This is most easily accomplished by limiting your fuel load. Just because you can go with full tanks doesn't mean you have to. Another trick is simply to go early or late in the day when it just is not so hot. (I am personally convinced that is the primary reason they invented daylight savings time!)

It is equally important that you realize that your engine thinks it is at a higher altitude as well and you must lean the engine for best power (typically about 125% rich of peak EGT) or max RPM (if you are flying a fixed pitch propeller aircraft) before you start the takeoff roll.

Remember that it is not just take off performance that is affected. Your landing distances increase as well. Even though you are flying the same approach speeds as indicated by the airspeed indicator, your true airspeed is significantly higher as is your ground speed, which translates into longer landing distances.

Also consider the ramifications of high-density altitude on your aircraft's service ceiling as well. I typically run my normally aspirated singles at 10,000 feet in the summer. I have seen density altitudes at that altitude in excess of 15,000 ft. That is higher than the service ceiling on many light singles. In short, density altitude is "performance altitude".

Another consideration is density altitudes relationship to your body. Not only does the aircraft "think" it is at a higher than indicated altitude, so does your body. While it is perfectly legal to fly all day at 10,000 feet indicated without the use of supplemental oxygen, you simply can't fool Mother Nature. That headache you experienced is not a sign that you need a new headset; it is a sign that you are hypoxic. Portable oxygen systems are so easy to use, so affordable and small in size that there is just no excuse for not owning and using one. I carry a little 9 cubic foot system that is so compact it fits in my flight case. With the use of nasal cannulas, it will provide two pilots O2 for nearly 8 hours at the altitudes I fly. That is roughly double the amount of fuel I have on board.

The F.A.A. has been watching density altitude related accidents for a long time. A few years back, at some of the airports that hosted more than their fair share of these events, the feds installed electronic density altitude signs at the end of the runway, advising the current density altitude to any pilot who cared to read it. I thought it was a great idea, they thought it was too expensive. Now, most automated weather systems like AWOS and ASOS give density altitude reports, and virtually every GPS has a density altitude computer built right into it, so there is no excuse for not knowing what the density altitude is.

When considering the effects on your aircraft's performance, error on the conservative side. Take the extra two minutes to break out the book and figure out the take off and landing distances as well as the expected climb performance. While you are at it, figure out your cruise performance as well. Optimum altitudes for normally aspirated aircraft are around 7,000 feet. If the aircraft thinks it at 14,000 feet you will find that it will be hard pressed to make even 55% of rated power.

The only truly great thing about high-density altitudes is that come next winter, you will simply marvel at the climb rate and cruise speeds your aircraft is capable of.

Pushing Your Personal Envelope

For all the things aviation is, one of the things it is not is forgiving. Among other things, aviation is not forgiving of bad judgment, it is not forgiving of incompetence and it is not forgiving of complacency.

As a general rule, pilots as a group do a good job when it comes to taking care of our equipment. If you look at the accident statistics, it is obvious that fatal accidents due to purely mechanical failure, beyond the control of the pilot, are few and far between.   The NTSB finds pilot error as a primary cause in the majority of fatal accidents. The single largest killer of private pilots is continued VFR into IMC, followed by loss of control. Though it could be argued that if those pilots had held instrument ratings, that statistic would reflect a lower accident rate, I would argue that it is more a reflection of poor judgment. Judgment is perhaps the one thing a flight instructor can't teach a pilot. If a pilot is reckless in the course of their everyday life, it is reasonable to assume that they will be reckless while in command of an aircraft.

Complacency and incompetence are things an instructor and a pilot can do something about. I know plenty of pilots who tell me they fly about 100 hours a year, and they are always current. But there is a big difference between currency and competency. Did they really fly 100 hours, or did they fly the same hour 100 times? It really does matter. The latter breeds complacency. You get comfortable in the aircraft, the airspace you fly in, and the mission profile you fly. So long as everything is fine, there is no problem.

I recently did a BFR for a pilot who flies a light twin. His logbook showed that he flew that plane, which he bought new in 1987 about 125 hours a year. It is maintained meticulously and looks new. It has all kinds of new avionics in it, including a new IFR approved GPS. This multi-engine rated private pilot held an instrument rating as well, and he told me he occasionally flies IFR, so as part of the BFR, I thought we would include some instrument work. Three approaches and three busts later, I began to get suspicious. During the VFR portion, I gave him a simple simulated engine out, in level flight, in the vicinity of the airport, and though he got it on the ground, he used every inch of that 5,000 ft runway. Not pretty. While taxiing back for takeoff I asked him when was the last time he had practiced a single engine approach and landing and he proudly told me, "on my last bi-annual!" To add to my discomfort level, he could not tell me how to use his shiny new GPS to shoot an approach, though the VFR "direct to" operation he had down to a science. I don't think he had turned on his ADF since the last BFR either.

This man was a classic example of a pilot who had grown complacent with his flight ability, and felt that lavishing his equipment with unlimited funds for improvement would somehow minimize his risk while in the aircraft.

Obviously, money wasn't the issue here. I suspect that time was his problem. He simply did not have the time to invest in remaining both current and competent in the aircraft.

So, what can a pilot do to remain both current and competent as an aviator?

Push your personal flight envelope at every opportunity. If you're not instrument rated, consider getting one. If nothing else, it will sharpen your VFR flying skills and raise your level of competency to a new high. Already have an instrument rating? Go out and practice things you never do. Go grab a buddy or better yet a CFII, as a safety pilot and go flying. When was the last time you practiced an NDB approach? Never mind that you would rather declare an emergency than shoot one for real. The simple act of going out and practicing it will force you to exercise long dormant flight and reasoning skills. Don't have an ADF, how current are you in shooting a GPS approach? How about a DME arc? A localizer back course? An NDB hold? If you have an autopilot, shoot each approach two times, once by hand and once coupled. Doing it in a twin? Do them with one engine out. Not multiengine rated? How about it? Even if you don't own a twin and don't plan to, getting back into the training cycle will make you a sharper, more competent pilot.

Even something as simple as going out to get your night currency will improve your flight competency.

A few years back I spent a weekend obtaining a seaplane rating. I haven't flown one since, but the experience was wonderful and I learned some things. I now have a new frame of reference when I read a story about seaplanes that allows me to appreciate it on a totally different level.

If you regularly fly off long hard surface runways, why don't you take an hour or so and go out and practice on a short grass strip. If your normal routine has you flying to and from uncontrolled airports, make it a point to fly into some busy class C or B airspace, just to scrape some rust off those skills.

In the FAR Part 135 world that I now live in, we as pilots must take a check ride every six months and recurrent training on each aircraft we are qualified to fly, every year. The fatal accident rate on 135 operations is significantly lower than for general aviation. But our flying is very different than typical G.A. flying. Very little of what we do is "routine". We regularly go to airports we have never seen before, often at night or in less than ideal weather. We file instrument flight plans almost all the time and request radar flight following when we go VFR. We regularly operate over water and into and out of short fields. Almost every hour is a quality flight hour.

A few years back I went to King Air school. Operating in the turbine world was full of new challenges. Operating in the flight levels, complete with ice in summer, triple digit wind speeds and much higher operating speeds than what I had experienced in the piston environment found me lying in bed at night with the ops manual, trying to get comfortable with ITT's, P3 pressures, hot starts and pressurization emergency procedures.

Ever wonder why the F.A.A. requires you to practice holds for your IFR currency when a real world hold is as rare as a winning lottery ticket? It is because the hold is a busy place. You need to do the mental gymnastics to get yourself into and out of the hold, plus adjust your speed and time to give you the leg lengths you want, while compensating for the wind's effect on your flight path. They figure if you can handle that mental exercise correctly, your situational awareness is sharp and your head is in the game.

And don't think that at some point you'll ever have enough flight time or experience to back off competency training. My partner is a twenty thousand hour former fighter pilot and former airline pilot. He is also a rotor wing CHII, and holds a U.S. Forest Service fire card and external sling load rating. Over lunch he mentioned that one of his buddies owed him a favor and he was going to take it in the form of a type rating in a Lockheed Constellation. That compound radial-engined triple tailed airliner was the last of the great piston liners before the jet age, and possibly one of the more useless type ratings I could think of. But this was no cocktail rating. This is the real deal. Four days of ground school before you ever get to touch the plane. It would probably cost ten grand if you had to pay for it. When I asked him why he would call in such a big marker on something like that he replied, " Because I'm running out of challenges in aviation." I should only be so lucky.

But I'll be the first to admit that my partner represents the minority. For the rest of us, we'll have to make do with what we have access to and can afford. It all comes from the commitment of the pilot. Whether you have 600 hours or 6,000, private pilot or CFII/MEI/ATP, there is always something new to learn or relearn. Stay in the training cycle. If the last time you rode with an instructor was on your last BFR, you are way overdue. Find one you like. Spend the time to let him or her learn your strengths and weaknesses. Don't be proud, ask the instructor to push you. You don't have to be a professional pilot to fly like one. All you need is the desire to expand your personal flight envelope.

Watching Good
Training in Action

The other night I had an interesting experience. After a long day of flying, we got back to our base airport long after dark. My chief pilot, who flies his Cessna 172 to work from his home on a private grass strip just seven miles from the one I live on agreed to fly me home and spare me the hour plus drive.

His airplane, a mid-1970's model 172, features a new engine and a Powerflow™ exhaust system. The rest of the plane is factory standard 172.

We did our run up, called the tower and departed. I commented on how nicely   that plane climbed with two on board. The tower cleared us out of the class "D" to the south. My buddy reached up to flip off the landing light and when he did the lights in the cockpit dimmed. I looked at the amp meter, which was showing full deflection discharge and snapped on the flashlight that I always have in my hand or near-by when flying at night.

I pointed out the anomaly. He went through the usual checks, resetting the alternator breaker, turning off and then back on the master switch. It only got worse.

So, here we are 2,000 feet over somewhere in the dark with a flashlight. The ceiling is 2,500 feet. We know this because we had just flown through it a half an hour before. There was no moon of any kind. It dawns on me that to most pilots, this would constitute at least a minor emergency. But the attitude in the cockpit was more of "I'm not in the mood for this s#%t" rather than of some impending doom. After a few minutes of unsuccessfully trying to get the electrical system back on line, my buddy turned on one radio and called Palm Beach Approach.

He advised them of our situation and that we were going to a private field west of town. He saw us as a primary radar target, thanked us for letting him know we were running around out there with no lights or transponder and asked if there was anything else he could do.

We told him no, and we would call him when we were in the pattern. He turned off the master switch to save whatever juice he could for the landing light.

This was not the first time this had happened to me. Eight years earlier, on a training flight with a commercial student, the voltage regulator in my Maule welded itself in the open position, and was continuously pumping 40 amps into the battery. I had my student turn off the master switch and fly home in the dark with a flashlight, only turning the power back on just before landing. If we hadn't done that, the battery would have melted down. In that aircraft, the battery is under the copilot's side (read; my) seat.

As we approached the private grass field in the dark, my buddy once again fired up the radio and contacted Palm Beach to let him know we had the runway in sight. Wouldn't you know that another aircraft arrived at the airport at exactly the same time, forcing us to maneuver some more in the dark and re-align ourselves with the runway.

The touchdown and taxi back were uneventful, except that the charging system came back to life, creating more frustration and speculation as to its source of the failure.

What I did not experience was any sort of stress or anxiety. We went through the failure procedures like we had done it a thousand times before. That was it! We had done it a thousand times before. It was exactly the type of failure we train for all the time. When it happened, we just fell back on our training and did what we were supposed to do. What I had witnessed was a textbook example of good training in action.

It turned out that the main bus cable had developed a break in itself sometime in it's 25 year lifetime, and chose that exact moment to decide to limit the number of electrons it was going to allow to pass through it. A simple failure that could neither be predicted nor prevented created a situation that required specific action. That is what all the training is for.

Lien On Me -
One Man's Adventures
In Aircraft Ownership.

Buying your own airplane, every pilots dream. But for those who haven't been there and done that, the best advise I can give you is Caveat Emptor. It is the things you don't know that you don't know, that can take what should be a wonderful life experience and turn it into the ride through bureaucratic hell.

I am talking about having clear title to your pride and joy. If you finance your dream, the lending institution will force you to run a title search to verify ownership. If you've done your homework you know what that is. A title search basically checks the records at the F.A.A. office in Oklahoma City to see if some person, institution or company has a legal claim against the asset (read aircraft).

If a person, bank or company is owed money, say for maintenance work performed, and they didn't get paid, they can place a lien on the property. That lien is registered with the F.A.A. The idea being that a prospective buyer would learn of this and if the seller wanted to sell his property, would be forced to satisfy the lien. Then a document called a lien release is generated, removing this stigma from the title of the aircraft.

While it sounds so simple, it rarely works that way. I'm no lawyer but I have had several interesting experiences with aircraft titles, of which the lessons learned are worth sharing.

Before we go further, realize that there are Federal laws, local laws and filing procedures vary from state to state so if my personal experience varies from one you may have had in your state, stay with me, because there is a moral to the story here.

About ten years ago I bought a little single-seat airplane from an estate. This factory built certified aircraft was built in 1952 and had been in and out of license for more than 20 years. The owner was a deceased airplane pilot and I bought the plane from a reputable, known dealer. He had a signed bill of sale from the deceased owner's widow, a woman herself in her late seventies.

Before I bought the plane, I ran a title search. The name on the registration matched the name on the bill of sale. There were no outstanding liens.

About three months after I sent in the bill of sale and new registration I get a letter from the F.A.A. that was basically a form letter. It said the signature did not match the signature they had on file from the previous owner who had sent in his registration.   Normally, that means that it was signed Mike instead of Michael, or the bill of sale included a middle initial or something.   Rule number one is that you must sign the bill of sale exactly as the name appears on the registration.

A call to OK City got a person on the phone. (This was in the days before the paperwork reduction act of 1997 and Al Gore's efficiency in government (oxymoron) program and someone actually answered the phone as opposed to now, where you simply get lost in endless voicemail streams, get fed up and hang up). They hadn't returned the bill of sale. Why, was my question. The answer was that they need another one with the correct signature on it.

I called the dealer, who put me in touch with the widow, who was kind enough to allow me to Fed-X to her another original bill of sale to be signed. Sixty bucks later it was on the way back to the Feds.

Three months later it comes back again, unacceptable. Again I get on the phone, this time to a supervisor. The signature doesn't match I am told. I explain that this is an estate sale and that the widow is probably signing the owner's name.

"Ah Ha! How do we know she has power of attorney!" is the reply. I don't know. So again I call this widow, who by now is starting to sound irritated with me. She gives me the name of her attorney. I call this man, who doesn't know me from Adam but agrees to send me a letter attesting to the widow's authorities and power of attorney over the estate of the deceased. That and sixty bucks for Fed-X has the proof on the way to the Feds and what I think will be the final resolution.

Three months go by (going on nine since I bought the plane) and I still have no permanent registration. But I have used this time to overhaul the engine and so I feel like it hasn't been a total waste of time.

One morning the mail pile brings me yet another rejection letter. A call to the supervisor to query "why" yields me an answer I would have never anticipated.

"How do we know he is her attorney?" Good question. "What would it take to convince you?" I asked. " Well, a letter from the dead guy would do it." There was this silent pregnant pause on the phone as I gathered my self-control and tried not to lose it. "Anything else?" I asked in a slow and deliberate tone that was intended to convey the fact that I was really not in the mood for sarcasm at this point. "No, not really" I was told.

At this moment I realized that legal intervention was required and I contacted an aviation attorney whom I know. Six months and five hundred dollars later, I had a good registration. All in all, a lot of grief and expense, for an airplane that cost less than ten thousand dollars.

The second episode in the wonderful world of aircraft titles was much more interesting.

I had my eye on this little twin that lived on my home field. I had watched it for years, knew the owners and had on more than one occasion let him know that I was a buyer if the price was right.   One day I was walking through the porch at the FBO and saw a for sale sign for the airplane on the bulletin-board, with a price of $24,000.00 I walked over to the owner's hangar and told him if he would take twenty, I was a buyer.

Months went by and I hadn't seen or heard a thing. One morning I came to work to find the plane parked on my ramp. A voice mail on the answering machine from the owner confirmed that if the offer was still good, the plane was mine.

Once again, I ran a title search. Nothing came up, so I did the deal for cash. Fast-forward four years. I had completely restored the little twin. When I was done, the plane featured new engines, props, radios, interior and paint. It was a ten. It was a fine flying example of a classic personal twin. Unfortunately, it was unusable for part 135 so I elected to sell it.

A buyer emerged ready to go....except on the title search he ran, two unregistered liens came up and his lender would not lend until they were resolved.

Unregistered liens are basically liens the Feds know exist but don't have much information about. Sure enough, his search turned up two unregistered liens that the search I ran four years earlier did not.

The first was an artisan lien dated 1980, from a now defunct FBO at an airport that had fallen to the bulldozers. The FBO's assets had been purchased by one of the big FBO chains in 1987.

The second was between the then owners of the aircraft. This husband and wife team had apparently been contemplating divorce, and each placed a lien on the aircraft to prevent the other from selling it.

First I retained the services of a title company. They found out that the FBO had a mechanics lien on the aircraft for $20,000, dated 1980. It was part of the assets purchased by a large FBO.

A call to the large FBO got me to the general manager. He was angry. Apparently, the acquisition of the smaller FBO nearly twenty years ago did not go smoothly. One of the sore points was this particular lien and he wasn't going to do squat to help me. Then he hung up.

Not to be deterred, I got on the phone to track down the husband and wife team. While I couldn't find the husband, I did find the wife, who apparently had a messy divorce, and was not pleased that I was calling to clean up this old mess from a previous life. She too would not take my calls after the initial call.

Now I was getting angry. By some incredible stroke of luck, the defunct FBO, the company that bought its assets and the husband and wife act had all resided in Palm Beach County, the same county I live in. So I got some advice from my attorney buddy who suggested that I go down to the courthouse and look up the cases for free.

I had no idea you could do that. 10 minutes on the computer down at the hall of records gave me a list of 13 documents I needed copies of, if I was to determine just exactly what had happened.

Unfortunately, it had happened so long ago that it wasn't on the computer, but on microfiche. How quaint. Twelve bucks and two weeks later, I had complete transcripts of both cases including the satisfaction of liens.

As it turned out, the big FBO sued for collection on the mechanics lien on the aircraft. The husband and wife, who were the legal owners at the time, settled the case for $8,000.00 back in 1987. As part of the settlement, they had to release their liens on the aircraft. The big FBO had accepted the $8,000 as payment in full and filed a satisfaction of lien with the court, but never filed a release with the F.A.A.

Out of the woods with the husband and wife situation, I now went back to the big FBO. I'll never know what happened in that deal more than twenty years ago but it must have been ugly. The CEO absolutely refused my call. Finally, I had to have my lawyer buddy send a letter, complete with court documents proving satisfaction of lien sent to the big FBO's chief counsel. Basically, it said if we don't have a lien release in 72 hours, we were going to sue for damages. It arrived at my office via messenger the next morning.

This process took me all of six months, and needless to say the original buyer for the plane had moved on. But I sold the plane in less than a week anyway.

Think this stuff only happens on old planes? Think again. I recently sold a very late model single (1997) for a buddy of mine. It had a lien on it for three times its value as part of a floor plan financing deal the dealer had with his bank. It had been resolved but the paperwork had not been completed. The dealer straightened it out in a few days but what if the dealer was out of business?

The solution? Title insurance. For a few hundred dollars you can buy a title insurance policy that guarantees clear title prior to the date you bought it. That doesn't protect you after you own it.

I know of at least one instance where a mechanic and an owner had a falling out over the cost of an engine change. The mechanic discounted the bill in order to collect from the owner then turned around and placed a lien on the plane for the balance.   There is no legal requirement to notify the owner that a lien had been placed on the property (at least not here in Florida) and the owner didn't find out until he went to sell the plane several years later.

Most states have pretty aggressive laws to protect small business owners from getting stiffed by unethical patrons, but not much protection for aircraft owners who are mixed up with unethical small business owners.

It is also important to get a sales contract or letter from the seller that holds you harmless for liens or claims placed on an aircraft prior to the date you bought it. Remember, different states have different rules, and you should always consult a qualified attorney for accurate advice.

The bottom-line; protect yourself from the things you don't know you don't know. Run a title search, get a good sales contract with a hold harmless clause, and if it is an expensive aircraft, spend the money for title insurance. Don't let a legal oversight ruin what should be one of the higher points in life, personal aircraft ownership.

No Damage History

It has been said that the three most common lies told are; The check is in the mail, the job is on the truck and the plane has no damage history. Those three words, no damage history, have tremendous consequences for the prospective and even current aircraft owners.

First of all, let me state that damage history is not necessarily a bad thing, if you know about it ahead of time. Sometimes you might hear it referred to as "known" damage history. If you know about it, and the repair has been done well, and the price has been adjusted appropriately, damage history is not a bad thing.

As an example, there is a late model low wing monoplane for sale here at my airport. It has 145 hours total time on it. It's air conditioned, loaded with radios and fancy autopilots and features a leather interior. New, that airplane sold for just over four hundred thousand dollars. In the first few hours, the new owner, a non-pilot learning to fly, and a flight instructor (from a major flight school), botched a landing. The damage included skin damage on the tail, a possible prop strike and nose gear damage. The repairs included new factory skins on the tail. The engine and prop both went to the respective factories for complete teardown, and the nose gear was completely replaced. Although there was no conclusive evidence of an actual prop strike, and the engine came back from the factory with a clean bill of health, the prop shop sold the owner a new set of blades for the prop. All re-assembled, that aircraft flew some 80 or 90 hours since the "incident" and as its only pilot since that time I can tell you that it runs beautifully. All the work orders and 337's are present in the logs. That plane is for sale for 30 to 35 thousand dollars less then would be if it had no damage history. The point is, for the savvy buyer, this plane is better than new, discounted for the damage and if you were to buy this plane and fly it for 10 years or until it was due for a new engine, the money you saved on the purchase would easily pay for the overhaul. You knew what you were doing going into the deal, and you bought it right. Owners know that damage history devalues an aircraft, so they are reluctant to elaborate beyond what is absolutely necessary when it comes to log book entries. Others are down right dishonest by simply omitting damage all together.

More often than not, damage history is not advertised. You travel some distance to check out your new potential bird, only to find out it has been on its belly, its back, in a hangar fire or some other disaster. But how would you know?

Rule number one; get someone experienced, someone you trust involved. Do not try to go out and assess an aircraft on your own! If you have seen the pictures, and have run a title search that has come up clean (the topic for another story) and you are ready to go take a look, get some help. You can save some time by getting photocopies of the recent log entries faxed or e-mailed to you along with the A.D. compliance list. Don't just get the engine and airframe either. Ask for the prop and accessories too. That's right, accessories. Everything from pneumatic door seals to wet compasses, fire extinguishers even radios can and have had A.D.'s issued against them. Sometimes, compliance can be expensive. Often, lack of compliance can make the aircraft un-airworthy thus illegal and uninsurable!

I can't tell you how many times I have heard horror stories about pilots who are just too proud to ask for help, or buy a plane on impulse. I can think of one man who bought a plane on his own, only to get it home and find out that the plane had been landed hard enough to waffle the skin on the fuselage. The ensuing repair cost him thousands of dollars. He had paid top dollar for this "clean and low time" aircraft. Now, a year or two later, he has had to top all the cylinders on the supposedly "new" engine. While talking with the owner, he expressed surprise that such a low time engine needed to be topped. (It was making metal.) Obviously, he was sold a bill of goods from the beginning, but it never occurred to him that his judgment (or lack of it) was the reason he was having all these problems with a supposedly good airplane.

When this man first bought the plane he brought it to me for some routine maintenance. That was when we discovered the waffled fuse. A look at the rest of the plane revealed rusted or frozen rod ends, un-checked corrosion everywhere, and questionable wiring. But most of all, the logbooks were just full of inconsistencies; missing entries, incorrect entries, and inconsistent compliance history. My IA at the time, also a friend of the aircraft owner walked into my office and basically said "I won't touch this plane with a 10-foot torque wrench!" There is no tactful way to tell someone they may have bought a pig in the poke, so we suggested he take the plane to a shop that specialized in that type of aircraft. Had I seen those logs before the sale, I might have wrestled the pen out of his hand as he tried to sign the check to buy it.

One of the more useful tools available in assessing an aircraft is a 337 search from the FAA. Every time a mechanic makes a major repair or alteration, he is supposed to fill out one of these forms. One copy goes to the F.A.A., one copy stays with the mechanic and the third goes in the logbooks for the aircraft. Occasionally, they get lost or separated from the aircraft records, but you can get a copy of every 337 sent to the F.A.A. You can request these online for a minimum fee.

Another thing I like to do is check the equipment list against what I see in the aircraft. When a piece of equipment is removed or replaced, the mechanic or shop doing the work needs to amend the equipment list and the weight and balance sheet. This is especially useful with older aircraft that have had radio changes. You check the changes on these documents against the logbook entries or 337's for the work. They should be consistent.

The next thing I do is look for obvious repairs, like skin replacement and check that against what's in the books. As I get more serious, I check the type certificate data sheet against what major components are on the airplane just to be sure the prop or engine in that plane, are certified for that plane. If not, walk away. I know what you're thinking....this guy has got to be kidding right?   Well I bought a brand new plane from the factory. I went to put it on a part 135 air taxi certificate. The maintenance inspector from the F.A.A. checked the serial number and model number of the accessories against the T.C. data sheet, only to find that the prop governor on the plane wasn't certificated for use on my model aircraft. A call to the factory got me a new one, and the realization that several other aircraft near my serial number went out the door with illegal governors on them as well. The F.A.A. inspector told me that often when a plane is damaged, parts will be replaced with new parts but few people bother to check to see if the supplied replacement part is legal. In this case the factory did it, but often it is done in the field.

Some damage is so common that the factory makes repair kits for it. Cessna wing tips are a good example. When you ding a high wing Cessna on the wing tip you wrinkle the outboard two feet of leading edge. The factory kit comes with new outboard leading edge skin, a new outboard nose rib and a wingtip butt rib. If you ding it a little harder, you may winkle the aft spar at the flap well. There is a factory kit to replace that as well. If you see these kits installed in an airplane you are looking at, it is because it has been damaged. The log entry may only say " Installed Cessna kit number ___ as per factory instructions," so you need to be able to read between the lines. Also if you see " removed and replaced" in the logs and they are talking about a major component like a landing gear leg, an elevator half or a prop, you need to ask " why?"

Missing logs are another problem. Are they missing because someone lost them, because a mechanic is holding them for lack of payment or because someone is trying to hide something? If it's the latter, a good physical inspection will catch it. If it is the second, a title search should address it. If it's the former, much can be reconstructed from the 337s and work orders. At certified repair stations work orders are like gold. They are required to keep them on file for a long time and they are very detailed.

Don't be afraid to call the mechanic, FBO or maintenance base that serviced the plane last. These guys make a living in aviation and few are willing to risk their license and lie about the condition of an aircraft, especially when they have little at risk.

Calling previous owners is also acceptable. I have owned several dozen aircraft in my life and from time to time I receive calls from people who are looking at buying an airplane I owned years ago. A few phone calls now can save you thousands of dollars later.

But still, you can be had. I wanted to buy a Cessna A-150 Aerobat I saw advertised. It featured 1,150 TTSN, and a low time engine. The owner faxed me log books, told me about some minor skin damage, which was in the logs, I ran the

337s and title search, and all looked O.K. That plane was in Midland, Texas, a six and a half hour airline trip including a change in Dallas. I took a young pilot with me since I expected to be flying it home to Florida. When we got to the plane I took one look at it and asked the owner to take me back to the airport. The skin damage reported in the logbook appeared as if had been repaired with air-conditioning duct aluminum and screen door pop rivets. To add insult to injury, it appeared to have been painted with a brush. There were several other repairs not in the book, rust or corrosion everywhere, automotive electrical connectors under the panel and every rod end was frozen. Not only was this plane not as advertised, it was downright illegal and unsafe. On the way back to the airport to catch a smoker home, the owner attempted to apologize, saying that he thought he represented the plane accurately etc.... He knew that aircraft was a "dawg", that was why he was selling it! That was $2,000 in airfare and a day of my time but well worth it in the scope of things. I'll bet ten grand wouldn't make that plane legal much less worth the asking price.

So while it is very possible that 'No Damage History" means exactly that, in the back of your mind that phrase should trigger a little voice that repeats over and over and over " Yeah, right."

Witness to Stupidity

If you hang around airports long enough and you are going to see things. Good things, amazing things and things you just can't believe you just saw. I have seen half a dozen aircraft accidents, two fatal and a few that left me shaking my head and wondering how I could nominate the pilot for the annual Aviation Darwin Award. For those who are not familiar with that particular award, it is generally given to the individual who could best benefit both the species and aviation by removing their genes from the gene pool.

Flash back two years. I am taking a Part 135 check ride with a Fed in a Piper Navajo. I knew this guy pretty well but he was giving me a good workout. We decided to stop at OBE, Okeechobee Airport for fuel. This uncontrolled field, at the north end of Lake Okeechobee is a busy training airport. All of the academy schools from Daytona Beach to Ft. Lauderdale use it for training, and its self serve fuel farm features Avgas for more than a dollar less than you can buy it anywhere else. Other than that, there is no reason to go there.

On final approach my Fed buddy turns to me and casually mentions how he hates to come to this airport because " You can always see someone doing something stupid." We circled to land on runway 4 and as we are taxiing in, a Cessna 337 Sky Master on final behind us pulled up into the flare. We watched as the airplane dropped in from about six feet up and catapulted back into the air.

" Uh- oh" I said directing his attention to the show on the runway. "Ohhhh!" I said as the pilot slammed the mix-master on to the runway again and again went airborne. This time the pilot made a mistake. Instead of adding some power and holding the nose up, on the next oscillation he pushed the nose over right at the top of the arc. Wham! The plane came down right on the nose wheel, knocking it clean off the airplane and sending it careening across the infield directly across the taxiway, in front of our plane. The airplane went up for yet another time and this time it came down on the mains, then the nose, skidding on the now un-wheeled nose fork, showering sparks as it went. You could hear the front prop striking the ground as the wounded Sky Master went past us on the way to the end of the runway. It probably skidded another 400 feet, before coming to rest perpendicular to, but directly on the centerline. The Cherokee behind it on final went around.

"Oh-Man!" my clearly pissed off Fed buddy cried into the boom mic on the headset. "Don't stop, take off so I don't have to get involved in this" he said. But the Sky Master blocked the runway, and we really needed the fuel. " Ya see! That's exactly why I hate coming here! Ahhhrggg!" he grabbed his head. "Ya know what, just taxi in to the FBO, I'm not gonna give the guy a hard time, I'm sure the damage to the plane is pain enough, but let me get his information since I'm here, that is if you have the time." "I'm in no hurry, I'll fuel the plane" I replied, figuring that I would be off the hook for the check ride.

We shut down in front of the FBO but there was a line for the fuel farm. Climbing out of the plane, the Fed flagged down the lineman, who agreed to run us out to the scene of the incident in his convertible Mustang. "Come on" the Fed said, "Jump in, ya never know if I'll need a witness." "Oh great I thought, what am I getting into now?"

As we pulled up to the Sky Master, the pilot was just getting out of the plane.

The Fed walked up and introduced himself by saying " Hi there, I'm from the F.A.A.!" and he flipped open his I.D. The expression on that pilots face was a Kodak moment if I ever saw one. Not only did he just wipe out his beautiful Sky Master (it really was exceptional) in front of all his friends, but as if on cue, out pops a Fed. I half expected him to say "and I'm here to help you", but instead there was a pregnant pause and I could see the Fed reading the dread on that pilots face.

"I'm not gonna give you a hard time, I just wanted to be sure everyone was allright, and get your information. Can I see your pilots certificate and medical?" If I was that pilot I probably would have just gone with the flow. But this guy goes the other way. He gets belligerent. "The gear collapsed" he said in an annoyed adamant tone. "Come on, I was in that Navajo on the taxiway. I saw the whole thing. The gear didn't fail" said the Fed in his best "I used to be a cop" voice. "Don't get excited, this isn't even my FSDO! I just want to get your information and I'll get out of here." The Fed was trying to calm the guy down. It was Friday afternoon at 4:30 and I'm sure he wanted to be somewhere else. "No, really, the gear collapsed, I didn't do anything wrong!" The pilot was starting to look and sound postal. He went on, "Ya know, I've got enough problems right now, I really don't need this..." and he turned away mumbling under his breath. To the Fed's credit, he remained composed. Here he is, thinking he's gonna cut this guy some slack, and the guy goes off on him. "Ya know what, I don't need this either. The gear failed? Yeah, right after you beat it against the runway a few times!" With that, he pulled out his cell phone and called the local FSDO, asked for the inspector on duty and reported the entire episode as he saw it. When he was done, he handed back the license and medical turned to the line guy who drove us out to the plane and said "Let's go." Then he turned back to the pilot and a few guys standing around the plane and said

"This is an accident site, do not move this airplane or touch it until the local F.A.A. office instructs you that it is o.k., do you understand?" Everyone nodded.

On the ride back to the FBO, the Fed said nothing. I'm sweating the rest of my check ride. While I fueled the plane he was on the phone with the local inspector on duty. I'm thinking, "Oh what a treat, a check ride with a pissed off Fed, lucky me."

We climbed back into the plane in silence and I started the engines, taxiing to the other runway. Finishing my run up I asked what was next. The Fed turned in his seat, looked at me and asked

" Was I wrong or was I getting an attitude from that guy?" "I don't know" I replied, "if it had been me, I'd be upset, but not with you, with me." I continued. Shaking his head he said "I told the guy I wasn't going to give him a hard time and he still went off on me. I mean, I was willing to let it go! It's too bad there is no FAR prohibiting stupidity, because if there was, then I could violate him!"   Interested in getting out of there, I asked again "Where to?" With that, he snapped back into check ride mode and we finished up a little after 6 pm.

Fast forward about a year. It had been raining hard in South Florida for five days. We set some monthly and daily rainfall records during the week. Myself and one of my line captains from my charter company, as well as and one of my line employees, an aerospace engineering student, are standing on the ramp at our home airport. It hadn't rained too much that day, but as we pushed the last plane into the hangar it was coming down pretty hard. We stood in the door of the hangar waiting for it to ease up so we could make a dash for the office and the cars when a Cessna 310, which had been on approach to runway 3 poured the power to it at about mid field and went around. At first I thought there was another plane on the runway, but it was raining hard, and as the 310 turned onto downwind, you could barely see it. The wind was clearly from about 270 at 15 to 20 knots and the aircraft that landed before the 310 had used runway 27. The windsock confirmed it. "That's about as minimum VFR as you can get" I said, referring to the indefinite ceiling and limited visibility as evidenced by the fact that you could hardly see the airplane, less than a mile away and well under 1,000 feet. "Doesn't he know he's downwind?" my engineering student asked out loud. "No way to know what he knows..." I answered. With that, the 310 came in for another attempt to land.....on the same runway! The aircraft touched down midfield. "Uh-oh, he's gonna try to stick it on the runway,.....baaaad idea! He's got too much speed, he's not gonna make it! Anybody want to bet money this is gonna look ugly?" I could not believe what I was seeing. Just as my captain was reaching into his pocket to take the bet, the 310 started turning sideways while continuing down the runway. " Uh-oh...." With that the gear folded, and the plane slid to a stop a few hundred feet from the threshold. " I told you it was gonna get ugly! Unbelievable!" With that, I picked up my cell phone and called the FBO office to report the incident. We certainly didn't want anyone else to land on that runway until they could get that mess cleared up.

"What happened" the engineering student asked sort of rhetorically, since he had seen what we had seen. "I'll tell ya" the captain chimed "he was going way too fast and downwind, touching down at the midway point in the runway. The trees at the end of the runway were getting bigger, so he stomped on the brakes, the wheels locked and hydroplaned on the standing water. The plane got sideways, and when it got to a dry spot, it knocked the gear out from under it." Yep, file that under stupid pilot tricks.

As I drove off the airport the fire trucks and rescue vehicles were on the way in. It made the 10 o'clock news. I felt bad for that pilot. He would be filling out paperwork for months, not to mention bending his airplane. But he really had no one to blame except himself. Landing downwind on a short runway in the driving rain, when an into-the-wind runway that was longer was available definitely falls under the category of pilot error.

The next morning my Fed buddy calls me on my cell phone. "Hey, you know anything about a 310 that folded the gear last night?" " Yeah, me and two of my guys stood there and saw the whole thing, why?" " Ah, just filling in the paperwork. You were standing right there?" "Yeah, right in the hangar door watching the whole thing. Had a front row seat, just like that time up at Okeechobee, remember that?" " Sure do! Well that makes two times for you. One more time and you could be considered a professional witness to stupidity!"

Things You Can Only
Learn In A Piper Cub

Back in 1989 when I bought my first Cub I had no idea how important that aircraft type would become to me in my flying life. Aircraft designs typically reflect the ideals and qualities that are near and dear to their designers. All airplanes are a series of tradeoffs; that is, you need to give up something to get something else. It became obvious to me almost immediately that Mr. Piper was willing to give up speed for docile handling, speed for ease of assembly, and speed for cost of production. That is to say that a Cub has been called many things....fast is not one of them. I did not know what to expect when I picked up my first Cub N3255M, a 1947 PA-12 Super Cruiser. By the time I bought my second Cub, a PA- 18 Super Cub I knew a lot more. When I purchased my third Cub, N3221N, a 1947 J-3C -85, I was deep into the world of Cub flying.   Selling that airplane was probably the single most regrettable thing I have done in my aviation life. My current (and most likely last) Cub is a 1947 PA-11-85. The PA-11 is basically a J-3 with a cowl like a Super Cub. Mine has an 18-gallon wing tank and a two-gallon header tank, plus an electrical system. The wing tank makes it possible to solo from the front seat, and the lack of the 12-gallon header tank means there is actually room for a pilot in the front seat. The electric system was a concession to reality. It still has a 1220 lb. gross weight. Other minor refinements include a bob-spring on the elevator reducing elevator pressure at slow speeds and the back of the front seat folds forward to allow easier access to the rear seat. It is also 15 or so mph faster than my old J-3 with the same engine. I would chalk that up to improved aerodynamics from the cowled engine vs. having the cylinders hanging out in the breeze.   When we are talking faster, we are talking relatively. A stock 65hp Cub goes 65mph +/-5, period. The 85h.p. engine improves climb significantly but cruise speeds only go up to 70-75mph. So the first thing you learn in a Cub is, well, patience.

Going cross-country from Morristown, N.J. where I bought my PA-12, on down to West Palm Beach, Florida was the first time I had been in a Cub. My instructor was a capable former banner tower from the Jersey Shore who had several thousand hours of tail wheel time. As we weaved our way southeast between Philly and Atlantic City toward Cape May at 1,000 feet I couldn't help but notice that this thing was S L O W! When I asked him if that was it, if that was as fast as it goes, he rotated in his seat to face me and with a big grin said "It's a Cub!" as if that was some universally understood pilot thing that I should already know. The second thing I noticed was that, once clear of the complex urban airspace, my instructor kept us at 1,000 feet. So again I asked. "You don't fly a Cub above one or two thousand feet?". O.K. I thought... this is a game show... "I'll bite...... the category is things only you can know for $100 and the question is WHY?" I said. " If they wanted you to go higher, they would have installed a mixture control knob" was the reply. Sure enough, I peered around his shoulder toward the throttle only to see.... no mixture control. I'd just file that one under things to resolve later.

Our first landing, my first ever in a Cub came at Easton, Maryland, a small single runway strip out on the Eastern Shore. I watched in amazement as the rudder pedals never once stopped moving from the time we flared till we pulled up in front of the fuel pumps. Learning tail wheel was going to be challenging. "How long is it gonna take me to be able to do that?" I asked my instructor, who at only 5' feet 6" tall was looking much larger than his actual size to me right about now. "Most people get it in 10 or 15 hours. Tt's really dependent on your basic flight skills." How right he was. At this point in my flying career I was a two hundred hour instrument pilot who had only flown 172s, 172 RGs and a Grumman AA1C. I did not realize that I was simply incapable of accessing my own basic flying skills.

So it went for the next three days, as we worked our way south, stopping every 90 minutes or so for fuel, avoiding congested airspace and following I-95. From my double-wide seat in the back (PA12's were certified 3 people, one in front two in back) I could spread out with my chart and follow the landmarks on the ground with finite accuracy. In fact, since I had routinely flown higher as opposed to lower, I had become good at picking out the shape of say a lake or a peninsula, but long ago gave up looking for railroad tracks, antennas smaller than 1,000 feet, and major roadway intersections. All of which were quite visible from my perch, here at 1,000 feet. Although the PA-12 had an electric system and a nav-com, (a big deal in a Cub) I don't remember ever using it except to announce our position over the Unicom frequency. I learned to fly in New Jersey at a tower-controlled airport and was accustomed to always talking on the radio. So cruising along in silence with the window open at 1,000 feet gave me the opportunity to notice things I never noticed in flight before.

As we worked our way through the Carolinas, you could smell the lilac, the pine and the occasional forest fire. As we approached the coast, you could smell the tidal flats and the paper mills. And over the din of the engine and the wind, through the headset and all, I could hear the fire engine on the highway one thousand feet below us. I even was able to read the names off a couple of water towers. If I wasn't so overwhelmed by the machine I might have learned right there and then the most important lesson I would ever learn in a Cub. But with 200 hours total time, I just didn't know what I didn't know, and so it would be many years later before that particular and all-important lesson would be finally delivered.

Eventually, I became comfortable with flying the airplane. In level flight the controls felt clumsy to me. A Cub simply won't turn without the use of coordinated rudder. Certainly not the little fighter plane my Grumman was. If you just try to stick it over, it sort of mush/yaws in the direction you want to go, but keeps going the way it was going. Obviously, Mr. Piper designed in a tremendous amount of adverse yaw. This came as a surprise to me, as my one and only experience with a Piper product prior to this was in a Cherokee. If you have ever flown a Cherokee you will agree that it is without a doubt the easiest aircraft in the world to fly. Mostly because the rudder and ailerons are interconnected and there is virtually no adverse yaw. You can land one in most conditions with your feet on the floor. Not so in a Cub. In fact, with time I discovered that the plane flies better if you lead the turn with rudder and coordinate with aileron as opposed to the other way around. So why all the adverse yaw in a Cub? Well, apparently Mr. Piper felt that if his plane was to be a successful training airplane, then it would need to teach pilots to fly coordinated flight. So while a Cub is an easy airplane to fly, it is a difficult airplane to fly well. Additionally, if you were trained in a Cherokee or as in my case a Grumman AA1, it is a good bet that your in-flight use of rudder skills are virtually non-existent.

I was convinced after just my short time with the machine that I would never be able to "feel" the ball out of center. In fact, the inclinometer, the "ball" may be the most important instrument on the panel. But from the front seat of that PA-12 I just couldn't "feel" anything. In fact, it would take teaching tail wheel in that plane from the back seat, to train my senses to the subtleties of a slip or a skid. It happened without warning, and now I feel it in everything I fly, tandem or side-by-side, without looking at the ball.

Learning Tail Wheel Landings in a Cub

Once I returned to my home base in Florida, it was time to start learning to master the take off and landings of a Cub. In the PA-12, you fly from the front seat, which affords a much better view of the runway environment than the J-3, which you need to solo from the back seat. Still, my personal learning curve was rather flat. It took me a long time simply to trust the aircraft not to stall when trimmed to 55 mph. I know (now) that it will fly much slower than that. But remember, being Grumman trained, from which a spin is reported to be unrecoverable, airspeed is a boy's best friend. And so, the first half dozen hours were wasted trying to land the thing way too fast. Again, a Grumman has a short fat wing, and a relatively high wing loading compared to the Cub. It would quit and sit down, where a Cub, flown into ground effect at 55 mph will float the entire length of a 3000 ft runway. Also, being of its vintage, a Cub doesn't have a stall horn (nor by the way does it need one), but again, my previous training just didn't allow me to accept that. I would become impatient and force the plane onto the runway, only to catapult back into the air like a cat that stepped on a stove. The bungee cord landing gear on a Cub can store a great deal of energy. When I finally allowed my instructor to slow the thing down to a reasonable speed (45-50 mph in the Super Cruiser worked great) trimmed to hands off (very important) with the throttle closed, the plane was easier to land. The reason I say trim is important is due to the design of the stabilizer. A Cub tail is attached at the front by a jack-screw, which is operated by the trim handle in the cockpit.   When you trim a Cub, you are trimming the entire tail up or down. If you don't trim properly, if you try to just muscle the nose up, which you can easily do, you never reach maximum critical angle of attack. You essentially "run out" of elevator. On the other hand, trimming basically full nose up makes it easy to get the nose up to just above the three point attitude, allowing the stall to occur at the maximum critical attitude instead of some lesser angle.

But again, I needed to learn patience. The Cub lands best in a full stall. That means that you need to hold it up, hold it up some more, and don't let the thing land till it drops out from under you. It took a while to get used to the high pitch angle as well. The little Grumman and Cherokee land fairly flat, especially with the use of flaps. On the Cub, with the nose held high, the forward visibility is blocked for the first couple of hundred feet down the runway, so learning to use my peripheral vision to stay aligned on the runway was a new skill I needed to master. Done right, it is a thing of beauty, the tail wheel kissing the ground a fraction of a second ahead of the mains. That is all well and good on a calm day, into the wind. Then we started working on crosswinds.

Crosswind Landings, Side Slips and Stalls

There are all sorts of stories about landing in crosswinds, and the ongoing debate as to the best method, the crab kick or forward slip, which is better? But the bottom line is, regardless of technique, you need to be straight with the runway before you touch down. Furthermore, you need to be tracking straight down the runway or you will ground-loop. Ground-looping was never a consideration in any of the trainers I had flown. Even the full castering nose wheel on my little Grumman only wanted to track straight ahead, no matter the wind. The Cub is very willing to tell you it is not happy with your interpretation of "aligned with the runway". Later on, as an instructor, I learned to evaluate private pilot BFR candidates with crosswind landing technique. If a

BFR candidate could handle a modest crosswind with finite control, usually the rest of the ride would be easy. But again, being trained in a modern design was a disadvantage in the Cub.

The Cub lands slowly, and is as forgiving as any tail wheel aircraft ever built. But it is simply not tolerant of poor crosswind technique. I found difficulty in many aspects of the crosswind landing. First, the Cub has no flaps, and typically, you are side slipping the airplane down to a position approximately a quarter mile from the touchdown zone and two to three hundred feet. So proficiency in sideslips in both directions must be achieved. Because you sit in the middle of the Cub, and can see equally well out of either side, you will always achieve maximum rate of descent from your sideslip if you turn the side of the plane in the direction the wind is coming from. That is, if you are lined up to land on runway 9 and the wind is from 150°, you would step on the left rudder, swinging the nose to the left and compensate with right aileron. When I first started slipping the Cub, my instructor asked why I had always slipped to the left. That is, right rudder, left aileron. I can only assume that being trained in a modern side by side trainer, we always slipped that way simply so that the pilot could see the runway out the left window while performing the maneuver. It never occurred to me to even try it the other way until he pointed it out to me. But being the skeptic that I am, I made him take me up and fly slips in both directions to the same runway starting at the same altitude at the same airspeed from over the same spot. A stop-watch and the VSI were the only tools I had (or needed) to prove or disprove this concept. But sure enough, after 8 or 10 passes, it was clear that turning the side of the airplane toward the prevailing wind got us down lower in the same time, or faster over the same distance. The only exception to that I was to learn later, is in a Cub with the clamshell door open. When you turn that big hole into the wind, it's like throwing out an anchor. My 85 h.p. J-3 had a VSI in it, and although there has got to be some instrument error present when slipping an aircraft with an open door, we could pin the needle. There is most definitely airspeed indication error when slipping. In my J-3 and PA-11, the airspeed indicates high when slipping. For that reason, and also to preclude the possibility of tail blanking, I slip the Cub at 60 mph indicated. That is way above stall, on purpose. Not that the Cub is mean in a stall. No, in fact, in a forward c.g. configuration, which in a J-3 is anytime there are two people in it, you can't...repeat, can't make it break. The stick will come full aft, the plane will mush along, ailerons working the entire time, descending about 600 feet per minute. There simply isn't enough elevator authority to make a clean break. The PA-12 is a little different machine. It is slightly heavier, 1750 lbs. Gross vs. 1220 lbs. for a J-3. It does however, share the same wing and tail area dimensions, thus yielding a higher wing loading. It also had a larger engine. Mine had the Lycoming 0-235, rated at 115 horsepower. While 30 horsepower hardly sounds like a big deal, it represented a 35% increase over my C-85 powered Cub and basically double the power of the original C-65 powered airplanes. It is interesting to note that they were all built under the same type certificate. From that I can only assume the CAA found little wrong with it. The PA-12 does behave much differently in a power on stall, especially a cross-controlled stall with power on, than any of its lighter cousins. When cross control stalling, it would not bite to kill, but it would break hard, tuck the left wing under and sort of quarter snap until the nose was well down and the airspeed increased. Fine if you are at 2000 feet during practice, but scary to the low time pilot, 200 feet and a quarter mile from the end of the runway. I even had the plane re-rigged by the local master rigger, who found little wrong with it. It still did it. Years later I flew another PA-12 which landed at our field on the way either to or from the show at Sun N Fun in Lakeland, Florida. When I asked the owner about it he confirmed the same experience, and absolutely refused to let me try it in his airplane. I could surmise from his reaction that he had been there and done that...and once was enough thank-you. So I decided to keep the speed up at 60 miles per hour when slipping the PA-12. The rest of the stall characteristics on that Cub and all the others I had were best described as virtually non-existent. A few years later, I was teaching an F.A.A. inspector the finer points of Cub flying. Being an ex-instructor, ex-airline type he was curious to see how the Cub stalled. He had never even flown in one before that day. I had him reduce the power to idle, and hold the nose up till the stick hit the stop. No break, full aileron control, 600 feet per minute down. With a dead serious voice he said "So go ahead and make it stall." "It is stalled," I replied. The expression on his face was like a kid who was just handed tickets to the World Series. Pleasant surprise. "You're kidding," he added, the grin now breaking into an infectious laugh. "Nope. Now want to amuse yourself further?" I asked, "Sure!" he said, still laughing "Keep the stick full aft, keep the ball in the middle with rudder and slowly bring up the power. Watch the VSI." At about 2,100 rpm, the VSI read zero. At 2,400 rpm, the most the engine would turn at that airspeed, we were climbing almost 500 feet per minute, indicating just under 35 mph, the stick still full aft. Needless to say that he had never seen that trick performed in any of the regional carrier type heavy iron he had spent all of his flying time in.

I learned that lesson on my own one day while taking off alone in a J-3. I started the takeoff roll with the stick full aft. But the wind was rather stiff that day, and I was slow getting the tail up. I was waiting for outside visual clues to my airspeed. I had not yet learned how to let the plane tell me when it is ready to fly. With the stick full aft, the plane just lifted... no, levitated off the runway and began to fly, totally scaring the hell out of me. I eased the stick forward ever so slowly, fearing I don't know what. So I took it up to altitude (2,000 feet is considered altitude) and played with that for a while. I needed to know if a gust of wind picked me up, or what. A few days later when the wind subsided, I went out to the place where I took the Cub to be alone and work things out. A 5,000 foot long, 300 foot wide, privately owned public use grass strip near Lake Okeechobee. Except for the occasional errant cow or horse on the runway, you harely ever saw another person or aircraft. That airport also featured self-service fuel, both Avgas and Mogas. My Cubs were all certified for both, so if I was enjoying myself and wanted to stay longer and play, or if there was significant wind (more that 12-15 kts is significant in a Cub from a fuel planning point of view) getting home would not be an issue. (More on this in another chapter). I repeated that levitation trick a dozen times from a standing start, and a dozen times it did exactly the same thing. Later, I would own a Maule, whose ancestry draws heavily on the Cub, and it performed in much the same way.

Teaching in the Cub

And so it went for several dozen hours. Just me and the Cub. Slowly I pushed myself to fly in more wind. Then in more of a cross wind. On and off the grass, and then on concrete. I would read stories of bush pilots who could take one of these things off a gravel bar in the middle of a river somewhere with a Moose strapped to the wing strut and think " maybe he could ...but not me...". Or read of the grizzled aviator who would cheat the severe crosswind by taking off diagonally across the runway to reduce the angle of crosswind. Slowly, I learned that the airplane was truly forgiving of poor judgment on behalf of its pilot, but was perfectly willing to point out when that pilot was lacking in technique. Flying the plane, getting it from here to there, on and off the runway was becoming routine. Doing it well was another story.

So I began teaching tail wheel in the Cub. Just as when I started teaching primary, and then instrument, I felt I had good enough command of the machine to teach in it, and I knew that teaching it would force me to fly it more precisely.

I would routinely take my charges up to my favorite grass strip for the initial few hours. Aircraft handling on the way up there (it's about 35 miles or a half hour or so), followed by straight ahead, full stall to a full stop landing on the grass. I'd start them in the front seat, transition them to the back after 5 or so hours. It was some of the most enjoyable dual instruction I have ever given.

One Florida winter morning I was flying with a woman I had flown with before. Though low time, she was a good pilot with good judgment and great hand-eye co-ordination. She was a great student and she was enjoying the Cub transition program. On this morning the temperature was about 68 degrees. A cold front had moved through and left some rain on the ground the night before but the sky was clear blue and visibilities were better than 10 miles. The wind was 330 at 15 when we departed the concrete runway of home base. The trip up to the practice field was slow due to the headwind, closer to 25 knots at our cruise altitude of 1,500 feet. This Cub, an 85 h.p. J-3, had the standard 12 gallon tank in the boot cowl. I figured it at 6 gallons per hour since there was no mixture control to allow me to lean it. So in reality, I'd better be on the ground in 2 hours ...period. I was sitting in back, enjoying the perfect morning when my student commented on the slow ground speed. In my mind I figured we would need to buy gas after our sixth or so landing, do a few more

114

and head back, with a stop at another county airport that has a paved runway, directly into the wind.

Six landings later, I suggested we stop after the next one for fuel and a cup of coffee. We taxied to the FBO trailer, next to which are the self-serve fuel pumps.

As we climbed out of the Cub I noticed the engine was shiny. Thinking I had an oil leak, I went to take a closer look. To my surprise, the entire induction system was encased in about 1/4" or clear ice. My student and I stood there, in total amazement at the phenomenon. The Cub's cylinders hang out in the breeze. But I never realized just how efficient a refrigerator the induction system on a Cub was until I saw that. Needless to say, I have never failed to employ the carb heat on a Cub regardless of the weather conditions, whenever the RPM's are below 2,000. But that turned out to be the least exciting part of my day.

The fuel pumps were in-op. No reason, just no power. So, off to the county airport with the paved runway we went. It was only 15 miles away....no problem. Upon arrival we discovered the FBO closed. In fact, there was not a soul on the airport.

At this point, I did not believe that we had enough fuel to return home. Running short on options, I decided to "borrow" some fuel from one of the tied down aircraft. Armed only with an empty soda can, I drained several gallons, one soda can at a time and poured it into the Cub. I did leave a note and a $10.00 bill in the cockpit of the aircraft from which I took the fuel. I certainly didn't want the guy to take off thinking his tanks were full but were in fact down two or three gallons. I did not sign the note. My student was mildly amused. I smelled like gasoline. Our return flight was uneventful and due to the wind, pretty quick. I was looking forward to pushing the plane into the hangar and going out for lunch, but I still had one more trick to perform.

Upon return to the pattern at home base we discovered that the Department of Airports had closed two of the three runways to do some maintenance. That left runway 15/33 as the active runway. Unfortunately, the wind had changed direction to about 060 at 25 gusting to 30. To make matters more interesting, the east side of that runway is lined with Australian pines, which were planted as wind breaks. They make an interesting wind eddy pattern over the runway when the wind blows hard. Furthermore, this being a training airport, the pattern was full.

The first two passes down the runway convinced me that we would either have to go somewhere else or land in the grass on runway 9. I said out loud to myself, "If we don't land on this approach, we will go somewhere else." But the limited fuel state of the Cub at that moment made me question the wisdom of that thought the moment after I had it. Then I thought about that grizzed old aviator who landed his Cub diagonal on the runway to reduce the crosswind angle.

On the third approach, I lined up with the left edge of the runway and aimed for the right side of the runway. Carrying a little power to make the rudder more effective I flew into ground effect, intending to wheel land the little Cub. Just as the left rudder hit the stop, the right aileron hit its stop and the right main touched on the pavement. For the longest 10 seconds I could remember, I did my best Bob Hoover impression. But we tracked diagonally across the runway, the tail stayed up, and as the left wheel settled, I killed the remaining power and let the tail down and nailed it to the ground. We were down and stopped in a few hundred feet. The ground speed being severely reduced by the now more head wind than crosswind. I easily made the mid field turn off. I couldn't have been more pleased with myself or the Cub. Lunch was uneventful.

A few months later, thinking I had learned that lesson, I got caught out with the same student and again couldn't find gas. I landed on a duster strip in the western part of the county, and was once again forced to "borrow" some fuel. My next Cub would have a wing tank.

The People You Meet

My J-3 Cub was a 1947 metal spar, 85 h.p. J-3. I bought it in Olive Branch, Mississippi during the summer months. It was straight, came with a metal prop (the wood one was on the plane) and it had all the logs. The plane had not been seriously wrecked...unusual for a nearly 50 year old plane. The fabric was sound but not pretty, but I didn't care. I had planned a full restoration in this plane's future.   Now all I had to do was get it home to South Florida. This J-3 had a 12 gallon tank in the boot cowl. I estimated the little 85 horsepower Continental consumed 6 gallons per hour. That meant a fuel stop every hour and a half. At a blazing ground speed of 65 miles per hour, I'd be stopping every hundred miles or so.

I was doing great. An early start and hand held radio got me 400 or so miles down the road before the afternoon boomers made flying impossible. The next day went equally well as I worked my way through Alabama toward the Gulf of Mexico. I was hoping to make north Florida. Getting airborne after yet another fuel stop, I was starting to get used to 5 miles in haze and sweating in flight even though the door was open. Approaching Eunice, Alabama, the weather started to turn deep south in the summertime nasty. I started to see lightening flashes through the gloom. It took me a few minutes to realize that with flight visibilities at about five miles, I couldn't be too far from that storm. The wake up call came when a cold blast of dry air came through that open door. Time to land. The nearest airport was just six or seven miles straight ahead. But straight ahead didn't look too promising.

I considered the weather east and west. There were airports 12 to 15 miles in either direction. Another lightning strike, this time close enough for me to hear the thunder. Time to consider a road, a field, a...... what's that? A private strip with a small pole barn at the edge of a farm field. I'm in. Just as I touch down, it dawns on me I'm miles from anywhere. The following thoughts run through my mind; the field is pretty rough, don't break this thing cause it's gonna be a long walk.

What happens if I ball it up and I can't walk? They probably won't find me till they plant the winter wheat. What about wind shear? Too late......I'm on the ground and rolling out toward the pole barn. The first big wet drops of the storm start to pelt the fabric just as I am pushing the little Cub backward into the pole barn. The front is wide open, but the storm is coming from the other way. Pure luck. I stand there pondering whether I should curl up inside the plane, close the door and close my eyes while this thing rains itself out or pull up one of the hay bales in the corner. Not knowing what creatures are living in the hay, I opt for choice number one. The rain is coming down hard on the steel roof and I can't hear myself think.

Visibility is maybe 100 feet. As I stare out the open end of the barn into the pouring rain, an old pick-up truck comes driving at me. I hear the theme song from deliverance start playing in my head and I prepared for a pissed off farmer with a shotgun.

Pulling right into the barn, I can see one man, with a young girl in the front seat. Rolling down the window he says with a thick southern drawl "Nice Cub. My daddy had a Cub just like it. Mind if I take a look?" "Not at all....err, this your strip? I got stuck in the weather and had no way out, saw the strip and the barn and figured I could get out of the rain......hope it's o.k..... didn't have anywhere else to land" I rambled on. " Soon as it stops I'll be out-a-here...." I continued rambling.

"My daddy used to dust the crops with his Cub. That's why we built the strip and the barn. Sold it years ago though. Always wanted to get another one someday.....that's why I kept the strip." The rain was torrential. I could barely hear him over the din of the rain on the steel roof. The wind had shifted and we were getting wet. "Gonna rain for a while," he said as he poked his head into the cockpit. "Any place to get some food round here?" I asked, not expecting an answer other than "No". "Well, the town is five, six miles up the road but we're headin home for dinner if you'd like to join us." Walking toward me he extended his hand and said "My name's Tom". "Michael, and thank-you but I couldn't impose... ." "Ah, nonsense. My wife would kill me if she knew I left a man out in the barn in the pouring rain....jump in." My city bred street sense just didn't know how to deal with this. I was cold, wet, tired, hungry and stuck in Mayberry. So off I went.

Dinner was excellent. The family was warm and friendly. We talked about airplanes, farming, life in the big city. It rained all night, and they extended their invitation to include spending the night. The accommodations far exceeded the prospect of a night in the Cub in a pole barn in the middle of nowhere. I fell asleep early. Good thing, because breakfast was at 4:30 a.m. Everybody was up and working, feeding the animals, doing laundry, cooking. When I got to the table at 5a.m. everyone was there and eating. I felt useless. There was no cable T.V. so we watched the A.M. farm report on the local PBS channel for weather information.

We talked more Cub talk. Tom told me how he would ride with his dad in the Cub when they sprayed the fields. How he loved the pull up at the end of the run followed by the wing-over and next pass down the field, over and over and over.

"More fun then a roller-coaster" he said. "Boy I miss that plane...." Looking at the far away look in his eyes I could guess that he was thinking of those special times he had spent in the Cub with his dad.   I've never been sure if he missed the plane or the plane just reminded him of his dad. "Time to get," he said, and everyone sort of jumped up at the same time. Five minutes later we were standing in the barn. The field was soft but useable. The air was thick, but there were no visible clouds in the pre-dawn light. "Does the airport in town sell gas?" I asked. "I'm not sure. Haven't been there in years. Why? Ya need gas?" "Well, I've got about six gallons, an hour maybe....." He cut me off in mid sentence "I've got a five gallon can in the back of the truck if you can take car gas."

He wouldn't accept a dime. Made me promise him a ride in my Cub. I said "How about right now" I offered, but he declined. He propped the engine for me, and waited till I ran up, taxied out and took off, waving as I went by. I'm sure a million memories were running through his head as he watched me climb away. I got a Christmas card from that family for years.

The whole episode was an eye opening experience for a New York boy who wouldn't expect a cab driver to slow down for a guy in a wheelchair trying to cross the street. I had heard of southern hospitality before. I'd like to think that man would have taken me home if he found me on the side of the road with a broken down Chevy. But I really think it was about the Cub. Somehow, my showing up on his strip in a Cub cued his long dormant memories of his time in his dad's plane and that made me ....well, allright. I have received a warm reception everywhere I touch down with my Cub. Seems anyone who has spent time in aviation has a soft spot for the little yellow machine.

Stepping up to a Twin

There has been a lot written about safety in multiengine aircraft. You've read about it, you understand that the twin can offer a quantum leap in capability and performance over most of its single engine brethren, if you're willing to put in the time to first become and then stay, a competent multiengine pilot.

So, what does it take to become a multiengine pilot? What does it cost to operate a twin? How about buying one?

Lets talk about training first. We have all seen the ads in the magazines for the guaranteed multi rating for $695.00. In New York, where I come from they would say "fegetaboudit!". Get real. Most of us know instinctively that you couldn't buy 5 hours in a twin for seven hundred bucks. Even if you did, what shape would the plane be in, how good is the instructor and just what exactly do you think you're going to learn in five hours! The ads also say guaranteed to pass your check ride. What they don't tell you is that that is a base package price and the average "student" takes considerably longer. As a multi engine instructor, I can tell you that the "average" multi student who comes to me with a private, instrument, and commercial rating and is instrument current, will take approximately 15 hours of dual and 6 hours of ground to get the commercial-multi-instrument ticket. That is an average. Some students take longer. One young guy, who was on his way to an airline career, came to me with the ink still fresh on his instrument commercial license. It took him 23 hours to get it right. He was a victim of less than adequate primary and instrument instruction, and those things needed to be fixed before he could safely fly a twin. I did have one student, who held only a private pilot certificate who did a VFR only multi rating and that took 9 1/2 hours of dual and five hours of ground for that student to master the airplane to the point where I felt comfortable sending him to the examiner. But a VFR multi engine private pilot is a rare thing these days. That is because the insurance companies are reluctant to insure a non-instrument rated pilot in a twin, and when they do (at a hefty premium) they usually require 50 hours of (dual) time in type before they let you go solo.

If you look at the Practical Test Standards (PTS) for multi engine airplanes you can get a good idea of what to expect for your training syllabus. Ground training will require that you have a through understanding of all of the systems on the aircraft, how they work, what to do when they don't. There are several important concepts that you need to comprehend if you intend to fly twins. Accelerate-stop and accelerate-go distances, balanced field lengths and Vmc are all concepts that the aspiring multi engine pilot needs to understand before you step into the airplane. Single engine operations need to be discussed thoroughly. When things go wrong in a twin most of the emergency checklists need to be committed to memory. The typical engine out drill in a piston twin is as follows:

Mixtures, Props, Throttles - Full Forward

Gear- up, Flaps-up, Boost pumps - ON (depending on engine type)

Identify- Dead foot, dead engine

Verify- Pull the power back on the suspect engine

Secure- Prop to feather, Mixture to idle cut-off, Mags off, Generator/Alternator off, Fuel off

Reduce the power on the good engine - No point in beating it to death.

Trim- If you are planning on flying any length of time, you will want to trim off the rudder and aileron pressures.

Each specific type has its own idiosyncrasies. For example, in a Cessna 337 Skymaster you leave the gear down until you reach a safe altitude, because the gear doors create a lot of drag when they sequence and the plane climbs better with one engine out if you leave them alone.

On my PA 23-160 Apache, it only had a hydraulic pump and it was on the left engine, so if the left engine failed the gear and flaps required that the pilot hand pump them up or down.

On my Twin Commander, the steering was hydraulic as well, and a left engine failure meant that you could expect limited nose wheel steering on touchdown. It did however have a good size hydraulic reservoir, which could be pressurized with the hand pump, something my Apache did not have.

Turbocharged Continental engines have two boost pump settings, high and low. Running the high pump with the mechanical pump functioning normally will result in engine failure.

Single engine work will include a Vmc demonstration. Vmc, or Velocity- Minimum Control is the speed at which the rudder can no longer compensate for the adverse yaw created by the asymmetrical thrust and the aircraft becomes uncontrollable. This is called Vmc rollover. The PTS calls for the demonstration to be conducted in a specific fashion. Typically, they want to see the left engine failed (on an aircraft with both engines turning clockwise), with that prop wind milling (Creating max drag on that side). On the right side they want max. manifold pressure and prop full forward (max power) creating the most possible asymmetrical thrust. The nose of the aircraft is then lifted to a pitch angle that results in a constantly decreasing airspeed, keeping the heading straight with the rudder, until the airspeed slows to the published Vmc speed, a stall is imminent or the rudder pedal hits the floor, whichever occurs first. Then the nose is pulled up ever so slightly so that the nose can no longer remain on the desired heading and the recovery is executed. Recovery requires that you lower the nose and reduce the power on the good engine simultaneously and return to the original heading. All of this is to be executed at a safe altitude, but low enough so that the "good" engine generates close to full power. I typically do this work above 4,000 feet above ground level.

Stall demonstrations in the multiengine PTS require only imminent stalls not full stalls. The relationship between stall speed and Vmc needs to be fully understood. When the Twin Comanche was first introduced it was used as a training aircraft. Its stall speed was close to the Vmc speed. Under certain situations like high-density altitudes the plane would reach stall before Vmc, setting up a dangerous spin situation. Some manufacturers began publishing Vssye or Velocity Safe Single Engine Speeds, to discourage going into that corner of the performance envelope.

You are required to feather an engine in flight and restart it, although through most of your training, engine failures will be simulated by applying whatever amount of manifold pressure is required to give you zero drag on the failed engine. In the Apache, that was 11 inches.

The PTS will require that you can execute a precision approach (read ILS) with one engine failed. You must also shoot two non-precision approaches (of the examiners choice) with one engine inoperative. If your aircraft is equipped with an ADF or a GPS you could well expect to be asked to demonstrate your skill with those types of approaches.

Landings with an engine in-op, simulated engine failure on take-off, and fuel management problems are all fair game on the check ride. On airplanes where it can be safely performed, manual gear and flap extensions with one engine failed is a possibility. What the examiner is looking for is your ability to deal with the emergency while maintaining directional control of the aircraft. Iron clad knowledge of the aircraft systems and theories of flight simply cannot be overlooked here and an examiner will see through any attempt to try to fake your way through the check ride.

If you find a good instructor with a good airplane, you will work hard, learn a lot and have a great time. Or if like me, I purchased my plane first and got my rating in it, using my training toward the 25 hours in type that my insurance company required before they would let me fly it as PIC.

What about ownership? What does it cost? I have owned half a dozen twins in the last 20 years so I think I can give you a pretty good idea. I'm sure you have read about formulas that are a multiple of the fuel burn to estimate operating expenses. I can attest that in general that theory works well. On a simple twin like my Apache or Beech Travel-Air with 4 cylinder engines, normally aspirated, you can expect to spend roughly three times the cost of the fuel at 16 gallons per hour. (It cost me $96.00 per hour to operate my aircraft. 16 gallons X $2.00 per gallon X 3 = 96) On a six cylinder, normally aspirated aircraft the fuel burn is typically in the 25 gallon per hour range or about $150/ hour and that was close to what my Twin Commander and Baron cost ($149 and $164, respectively). Step up to a turbocharged airplane like a Seneca or Turbo-Skymaster and you are looking at closer to four times the fuel cost per hour. The same goes for geared engines. My Twin Bonanza with Lycoming GO-480s burned 35 gallons per hour and it cost us right about $275 per hour to run it. Add pressurization, and it runs closer to five times the fuel cost. Add geared engines and pressurization like on a Cessna 421 and it can go higher than that.

My numbers include things like insurance, recurrent training and engine reserve. If you sell the plane before you run out the engines you'll never spend that reserve money, but the market tends to compensate for high engine times with lower sale prices.

Now obviously, certain fixed costs like hangar, insurance and recurrent training go down on an hourly basis when you fly more hours per year. But the average twin owner is flying just under 200 hours a year.

Thinking about buying one? Well, consider this; if an engine AD comes out on your engine you will have to do it on both engines.   A few years back Continental had an emergency crankshaft AD come out on new IO-550's and those that were factory overhauled. The A.D. required disassembling the engines. If you happened to be a Baron owner, you were not only down for the inspection but out the cost to do it....times 2.

When twin shopping, be suspicious of dual engine changes at the same time. Most twin owners like to change one engine, break it in then change the other engine. The reason for that is if a new engine is going to fail it is generally accepted that it will do so (catastrophically) in the first 100 hours. So you would want a known commodity in the other engine to count on. A double engine change can be an indication of a double prop strike or gear -up.   That is not always the case, but something to consider. Most twins are avionics heavy. Things like radar, radar altimeters, sophisticated autopilots and dual flight instruments, all desirable in twins, are expensive to repair and maintain. De-icing boots, pressurization, heated windshields prop anti-ice and gas fired heaters are expensive to repair, and some systems have recurring AD's which need to be addressed.

Like all aircraft, twins have type specific maintenance issues that you need to consider.

PA-23 series aircraft use fuel bags and they are expensive to replace if they are old and begin to leak. PA-34 series aircraft as well as Twin Comanches have landing gear trunion issues, which are costly to repair. Some Navajos have mandatory spar inspections and tail spar replacement A.D.'s which are recurring. My Twin Commander had a mandatory main spar X-ray inspection that had to happen every 36 months. It cost $4,500 to comply. The same aircraft had a one time flap pulley A.D.... cost $1,200.00. Hartzel had a major A.D. on many props commonly found on light to medium twins. On my Apache it cost $8,000 to get legal props that still required an inspection every 500 hours, but it was better than the $16,000 replacement cost for new props with no A.D. On turbocharged twin Cessnas the exhaust systems have multiple A.D.s out on them.

And while I hate to be a skeptic, just because it says it was done in the logbook, be sure your mechanic checks to see that it has been done on the airplane. I bought an airplane years ago (a Cub) that had all A.D.'s complied with in the logs. When I had a shop perform an annual inspection on it several months after I bought it, I discovered NONE of the work was done on the airplane. When I confronted the previous owner and his mechanic they basically knew they had gotten over on me. They actually thought it was funny. While a letter from the FSDO and my attorney wiped the smile from their faces, it didn't get my plane fixed. What really upset me was that they had put my family and me at risk just to make a few bucks. Don't be lazy, do the research. Call the previous owner, his mechanic, the engine shop the prop shop; ask to see copies of the 8130s on overhauled accessories. If the books are suspect, walk away. If the books are perfect but the plane is suspect, walk away. I can tell you that I look at 10 or more aircraft for every one I buy. I have on more than one occasion bought a run out aircraft and restored everything. That way at least I know what I have.

I bought my Apache in 1995 for $20,000. It was slap wore out. I had both engines overhauled, both props and governors overhauled and brought up to comply with the A.D.s, New paint and interior, brakes, radios, glass, cables, hoses, rebuilt the hydraulic power pack and put all new o-rings in the actuators. I overhauled or replaced most of the instruments. It took nine months and another $24,000. When I was done, I had a zero timed 5 seat multiengine airplane that had a 1600 lbs useful load, flew at 140 knots with multiengine redundancy on 16 gallons per hour. All that for less than fifty thousand dollars, and I knew what I had.

Flying a twin definitely gives me some peace of mind when coming home from the islands at night, over water or coming across the big swamp they have down here they call the Everglades. Especially when I have my family in the plane. A few years ago a guy in a Comanche 260 had an engine failure at night over the Glades and successfully put the plane down. He spent the night in the plane fending off the gators and snakes, and the mosquitos had a night on the town with this man. But when they pulled him out of there the next day he was none the worse for wear and he bought a Seneca with the insurance money.

If you've read the stories I've written on multiengine safety, I have emphasized the necessity to be realistic about your ability to make the commitment to the requisite training to remain both current and competent. The same advice holds true for twin-engine aircraft ownership. Be realistic about the mission profile you need to fly, and buy the equipment that meets that need. Be real about your ability to pay for it and you will be rewarded with great ownership experience. If nothing else, when you pull up in front of the FBO at the big airport in a twin you'll have a good chance at getting the carpet instead of being parked across the ramp with the other "little" airplanes.

The Realities of Flying
a Multiengine Aircraft-

Back in 1927, when a then unknown aviator named Lindbergh was asked why he chose a single engine airplane for his Atlantic crossing attempt as opposed to a multi-engined ship like many of his rivals, he replied that a multi-engine ship simply doubled his chances of experiencing an engine failure.

That piece of logic synopsizes the reality of operating a twin in a nutshell. Although engine technology has improved greatly, the law of averages hasn't changed since 1927. When you fly a twin, you have a greater chance of mechanical failure, and that is where the safety issues begin. There are lots of arguments as to the wisdom of that, because there are as many disadvantages as advantages in operating a twin engine airplane.

So why would anyone want to own one? What are the pitfalls? The advantages? The statistical realities?

Lets start with who. Anyone who routinely fly's over inhospitable terrain comes to mind. I for one, would be much happier flying over water in a twin then in a single. The same holds true for mountainous terrain, where a safe off field landing is unlikely. What about at night? Or more specifically, IFR at night. Sure, you can operate a single over water and mountains; it's done all the time. But a serious mechanical failure means you MUST make an off field landing. Statistics show that most people survive a ditching at sea, and an off field landing in the mountains, only to succumb to exposure or drown. In a multiengine aircraft, an in flight engine failure is still an emergency, but you have more options. Those options, not available to the single engine pilot, are precisely the thing that gets multiengine pilots in trouble.

A perfect example of that decision process presents itself to me as I decide how I want to go to the AOPA convention in Long Beach California from our base in West Palm Beach Florida. I own a fast single and a cabin class twin with an identical top speed as my single.

The trip is about an hour shorter in the twin because I can fly over the Gulf of Mexico from Sarasota Florida to just west of New Orleans. Plus, I could make the trip in one day (it be a long one) with the final leg occurring at night, but over the mountains.

In the single, the over water portion is out (at least for me) and over the mountains at night is also out. The difference (on this trip) is in dollars. My single engine Mooney M-20J, burns 11 gallons per hour, and the overland routing will take 13 hours and 45 minutes of flying on approx. 145 gallons, plus three or four fuel stops.

The twin on the other hand a Piper Navajo, using direct routing will take 13 hours @ 33 gph or 429 gallons, about three times the fuel burn of the single.

The other big advantage to the twin on this particular flight is the stuff I have to take with me. Two passengers and myself are no problem in the single or the twin. But, I also need to take material for the trade show I'm attending and the booth itself. If I go in the single, that stuff just wont fit along with a week's worth of luggage for three so I'd have to ship that material to and from the show, with some cost attached to that.

So what it boils down to is this; your personal mission profile. Mine has changed greatly over time. Ten years ago I owned a large Advertising Agency and I was flying photo crews around the country shooting catalogs so I needed a big twin. I bought a Twin Commander. When that deal was done I sold the Commander and bought a fast single because my mission profile was basically me and one or two other people, and less than 1,000 NM trips, day IFR.

If you fly for business vs. personal or pleasure use, where trip completion on a schedule becomes more important, or if you have the need to fly hard IFR, or IFR at night, or if your mission profile finds you wanting to fly four or more really large people and stuff on a regular basis, you need to consider a twin.

But if you are feeling the need to go multiengine there are several safety realities you want to know about.

Where would you expect to find the greatest number of accidents in a twin? As an instructor I tell my students that the takeoff phase is the most dangerous phase of flight. Low and slow with the gear down, an engine failure on takeoff would present the most problems to a pilot. But the statistics just don't support that logic entirely.

According to the Nall Report prepared by the AOPA Aviation Safety Foundation, nearly 34% of all accidents in multiengine aircraft occur in the landing phase of flight, as opposed to just 21% for the take off phase. It is further interesting to note that fully 25% of the fatal accidents in twins occurs in the approach phase of flight vs. just 5.4% in landing phase and 18% in take off phase.

So what does all that mean? Well, it basically means that a takeoff accident is most likely going to be fatal while a landing accident is not, but you have a far greater chance of being involved in a landing accident.

Also realize that multiengine aircraft represent only about 7.8% of the total GA fleet, but in the period studied in the 2002 Nall report, multiengine piston airplanes were involved in 9.4% of the total accidents, and 14% of the fatal accidents.

When and where these aircraft are operated also has an effect on the statistics. People buy twins for their greater capabilities. These capabilities include the ability to operate in more weather then the average piston single. Interestingly though, only 16.2 % of the fatal accidents occurred due to weather related issues in twins vs. 47.2% in retractable singles. So the statistics support the theory that twins are in fact safer in weather then singles.

Night also plays a significant role in accidents. According to the report, 6.9% of all accidents occur at night, and 66% of the instrument approach accidents occurred at night. Unfortunately, no breakdown of multi vs. single engine aircraft is provided.

A look at the FAA published Annual Review of Aviation Safety for 1996 (this was publish in 1999 and is apparently the last year for which complete statistics are available) clearly shows that the fatal accident rate in twins is consistently nearly double the fatal accident rate in singles.

So what makes a twin so dangerous? Most twins utilize nacelle-mounted engines on the wings. (We will exclude centerline thrust twins for the purposes of this conversation). This design comes with an inherent flaw, and that is asymmetrical thrust. Counter rotating propellers reduce the amount of asymmetrical thrust by moving both thrust lines closer to the centerline of the aircraft, thus eliminating the so called "critical" engine. (Despite what you may have heard the critical engine is <u>not</u> the engine that is still running after an engine failure, it is the left engine on an airplane with conventional rotating propellers, or the engine which has the thrust line closest to the centerline. When that engine fails, the remaining engine, the engine which the airplane is flying on, has the thrust line farthest from the centerline of the aircraft thus creating the larger amount of asymmetrical thrust.) That is why you see boom mounted engines on the tails of jets, because they are so close to the centerline that failure of either engine will not create significant amounts of yaw. In a typical light twin like a Cessna 310 or an Aztec, failure of the critical engine will create significant asymmetrical thrust, requiring the pilot to initiate the standard drill for engine failure in a twin.

Though there are variations for specific types the basic drill is as follows;

Mixtures, Props, Throttles- Full Forward
Gear up, Flaps up, Boost pumps –ON
Identify-Dead Foot = dead engine
Verify-pull throttle back to idle on suspected engine
Secure- Pull the mixture to idle cut off on dead engine
Feather- feather the prop on dead engine to reduce the yaw

It sounds simple enough, and in practice, it happens just about as fast as you can say it. In some situations, like take-off, the mixtures props and throttles are already forward, the boost pumps should already be on and that leaves the gear and flaps. Some aircraft models specify takeoff with partial flaps extended while others do not. That leaves the gear. Almost all light twins climb better with the gear up than with the gear out. When I teach multiengine departures, I want to see the aircraft's nose wheel on the ground until Vmc, and rotation at Vmc plus five. When we establish positive rate of climb, gear up immediately while simultaneously accelerating to blue line.

This narrows the window of time in which an engine failure would make the aircraft uncontrollable (Vmc roll over). By keeping the nose wheel on the ground to Vmc, you insure that you can maintain directional control until such time that there is sufficient rudder to do the job. By rotating at Vmc plus five, if the engine where to pack it in at the exact instant you rotated, I advocate killing the power on the good engine and stopping on the runway. It would be better to run off the end of a too short runway at 20 kts then to try to fly it off so close to Vmc. Why do I say that? Because if you look at the fatal accidents that have occurred on takeoff in twins, almost all of them are Vmc roll over accidents. It is obvious that the pilots of these airplanes elected to continue the take-off even after the engine failed. I witnessed that exact accident scenario right here at my home airport.

A doctor, who owned a P-Navajo elected to continue a take-off even after failing to achieve anything like take off power on one engine. He got less than a mile from the airport and crashed upside down killing himself and his passenger. Witnesses, myself included, heard him run-up and take off, the engines were never in sync. Had he killed the power on the good engine, he probably would not have been able to stop on the remaining runway, and probably would have wiped out his landing gear, but he and his passenger would have walked away for sure.

The bottom-line? Multiengine piston aircraft do not perform well on one engine. When an engine quits on a twin, you don't loose half the performance, you loose upwards of 80% of its performance. Furthermore, there is no certification requirement for these aircraft to climb on one engine under any specific conditions.

So when you read single engine climb figures quoted for specific aircraft remember that those numbers where achieved by a factory test pilot, in a factory new aircraft with factory new engines in what should be considered factory test air. (59°F, low humidity and 29.92" of mercury at seal level). Also remember that the test pilot knows the engine is going to fail and is spring loaded to respond.

When it happens to <u>you</u> for real, there will be several seconds of threat recognition and resolution that will pass before you actually react to the engine failure.

Furthermore, even if you do everything right, you will be rewarded with a climb rate of not more than a few hundred feet per minute, about half of what you might expect from a fully load and very tired Cessna 150. The difference is that in the 150, you achieve that rate of climb at a speed of about 68 KTS vs. 95 to 120 KTS in most light twins. You are covering a lot more ground in the twin.

Here are some published best single-engine rate of climb figures for popular twins;

Cessna:

MODEL	PUBLISH SINGLE ENGINE RATE OF CLIMB
310R	370 fpm
340	250 fpm
402B	225 fpm
414	240 fpm
421B	250 fpm

Piper

MODEL	PUBLISHED SINGLE ENGINE RATE OF CLIMB
PA 23-160	180 fpm
PA 23-250	235 fpm
PA 30 B	260 fpm
PA 44	180 fpm
PA 34C/R	225 fpm
PA 31-310	245 fpm
PA 31 P 350	255 fpm

As you can see, none of the models offer what could be considered a wide margin of performance. Remember these are the factory numbers. It is doubtful that you could achieve performance this good in a typical 25-year-old example with mid time engines.

Now lets talk about specific aircraft types and accident causes. I have chosen the Cessna 310 and Piper PA 23-250 Aztec (direct drive engine, un-pressurized cabin), the PA 34T Seneca (turbo-charged direct drive engine, un-pressurized cabin), the PA 31 Navajo (direct drive, turbo-charged engine, un-pressurized and pressurized cabin) the Cessna 414 (direct drive engine, pressurized cabin) and the Cessna 421 (geared engine, pressurized cabin) As a representative sample. That is not to imply that Twin Commanders, Barons or Aerostars are not worthy of discussion, but the aircraft I selected for the story represent a statistically large proportion of the multi-engine fleet. We will look at accident trends from January 1996 through September 2000. Some of the NTSB reports were not final at the time of this writing, but the preliminary reports are in.

The Cessna 310

There were 82 accidents according to the NTSB involving 310's from 1/1/96 through 11/1/00. Some of the investigations were not final, but there are some significant trends to be observed here. There were 11 landing gear failures during that time period. Some the result of hard landings, some the result of poor maintenance, but none were fatal. In that time, 4 pilots simply forgot to put the gear down, or lost it on the landing roll out. Again, none of those were fatal.

There were six reported cases of mechanical failures. One pilots seat collapsed on take-off, one pilot left the oil cap off of one engine ultimately resulting in an engine failure, one pilot left the tow bar attached to the nose gear and went flying, only to jam the tow bar in the gear well upon gear retraction. One pilot ran off the runway after the brakes completely failed and one pilot experienced an engine failure for no determinable reason. One pilot lost his life when the elevator control failed for undetermined reasons. There were three training accidents, two of which were fatal, one, a mid-air, and one that can be described as CFIT in VFR. It would appear that both the instructor and pilot/student had their heads inside the cockpit and flew the plane into the ground. There were 3 midair collisions involving 310's, all were fatal. One, a Part 135 air taxi flight being conducted under IFR flight rules, collided with a single engine aircraft, just outside of class D airspace as it was climbing through 2,500 feet. The single was not talking to anyone at the time. The majority of the fatal accidents involving 310's were determined to be pilot error during IFR. Some accidents occurred on approach, several pilots going below minimums, some were loss of control accidents while maneuvering, but more than half involved and engine failure or engine problem, and night IFR was a factor in three of those incidents.

The Cessna 414
23 accidents were recorded in the 414 during the period 1/1/96 through 11/1/00, nine of which were fatal. VFR into

IMC accounted for two incidents, both were fatal, 4 engine failure reports, (including one complete failure and a partial failure on the other engine) resulted in no fatalities. One stall on final approach killed the pilot and passengers, and a primary gyro failure on a particularly dark and foggy night also accounted for a fatal accident. One fatal occurred when ground personnel walked into a spinning prop, killing him instantly. A Part 135 training accident took the lives of three crew, including the chief pilot, of an air taxi operation.

More interesting, were two cabin pressurization failures. Although this type of accident is considered rare, it was brought into the spotlight a year or so ago when golf pro Payne Stewart's Lear flew halfway across the country by itself, crew and passengers unconscious before crashing when it ultimately ran out of fuel. A similar accident in a King Air 200 occurred in Australia earlier this year. For those who would consider pressurized equipment understand that this can and does happen. Two 414's experienced cabin pressurization failure, one incident was fatal, in the other the pilot managed to get the aircraft on the ground despite his hypoxic condition.

Some other interesting accidents involve ice as a factor. One 414 crashed on landing because the pilot could not see out of the windshield. Although the aircraft was equipped with alcohol windshields, electrically heated windshields are required for that aircraft's know ice certification. The last fatal accident in the selected time period occurred in Mexico on a Part 135 flight, where the crew flew the aircraft into the ground (CFIT) on a dark night in the mountains. Mechanical failures were few. A brake failure caused one pilot to depart the runway on landing and a gear collapse brought another pilot to grief.

The Cessna 421

There were 30 accidents reported during the period in Cessna 421's. Ten of those were fatal. One of the more interesting incidents that resulted in fatalities was of one pilot who took off with a known oil leak on one engine, which ultimately resulted in failure of that engine. The flight was IMC and in ice at the time and the pilot ended up loosing it and killing all on board. An air taxi incident in France resulted in 8 fatalities when the pilot lost an engine on takeoff. One pilot, after filling the fuel tanks with water on purpose, (to look for a fuel leak), took off and experienced double engine failure. Another pilot and passenger died when the pilot aborted the take off roll too late and ran off the end of the runway. His cargo, mostly medical gas cylinders were unsecured and neither occupant was found wearing a seat belt. They died from injuries inflicted by the loose cargo running around the airplane during the over run.

One fatal accident was the classic Vmc roll over from an engine failure on take-off. Another engine failure on take-off left the aircraft unable to climb and the pilot crashed into a shopping mall a mile from the departure end of the runway. One CFIT accident occurred at night and one piston failure led to loss of control in VMC. Both pilots perished in those accidents.

Some of the non-fatal accidents could have occurred in any aircraft. One pilot hit a deer on the runway, one pilot while exiting the pilot's seat to close the door hit the throttles with his foot and the aircraft ran into a hangar (the engines were <u>running</u> at the time). One interesting incident had two pilots flying a Part 135 trip. Although only one was required by regulation, the client requested two pilots. Both were involved with the pre-flight inspection of the aircraft. On take-off, the nose baggage door popped open and the flying pilot elected to abort damaging the aircraft. Sighted as primary cause was the lack of CRM training, which you might anticipate, in a single pilot environment. Other mechanical failures included total electrical failure at night, a trim tab failure and an improperly installed starter motor, which led to engine failure, and water in the hydraulic system that led to brake failure. One pilot stalled the aircraft on approach to final while heavily iced with inoperative de-ice boots. Unknown to the pilot the boots were badly holed. There were few purely pilot error incidents, the most egregious being a pilot who flew the airplane completely out of gas and crashed while trying to stretch the final.

This occurred while IMC and the plane departed into a spin. All aboard perished. Another pilot lined up with the runway edge lights instead of the centerline on a snow covered runway, wiping out the landing gear on one side while another pilot on a training flight tried to depart in deep snow and slush and aborted too late, running off the runway.

The PA 23-250 AZTEC

There were 40 accidents according to the NTSB involving Aztec's from 1/1/96 through 11/1/00. Some of the investigations were not final, but there are some significant trends to be observed here. There were 6 incidents directly related to poor or inadequate maintenance. One runaway electric elevator trim incident was fatal. Three pilots experienced complete to partial engine failure due to contamination of the fuel from corroded gas-collators, fuel screens or other fuel system parts. One gear collapsed due to bolt failure and one pilot lost control on landing due to brake failure.

Ten pilots came to grief due to whether. More than one pilot crashed while on approach because he elected to go below MDA or decision height without sighting the runway. Two incidents were attributed to low-level wind shear, and one was fatal. Ice played a factor in five accidents, one was fatal. By far the single largest number of incident were caused by loss of control following and engine failure. Six incidents were a direct result of the pilot's inability to maintain control following an engine failure in PA23-250's. The scariest incident was a blown cylinder, which resulted in an engine fire, which quickly developed into a wing fire. The event started at 6,000 feet, and the aircraft suffered wing separation before the pilot could get it on the ground. All on board perished.

One pilot flew his Aztec into the ground to avoid hitting a flock of birds. Yet another mushed into the runway while attempting a single engine go around in actual IMC.

Another fatal occurred when a low time pilot experienced a baggage door opening on take-off, leading to complete loss of control. And last, a pilot wrecked his plane while attempting to land after a passenger door popped open in severe low-level wind shear. Two of the recorded fatal incidents occurred in foreign countries and the details of those events are just not complete.

The PA-34 Seneca

66 accidents were recorded in the Seneca during the period 1/1/96 through 11/1/00, 16 of which were fatal. Interestingly, in only one instance was VFR into IMC attributed as the primary cause of an accident, but it was fatal. 3 engine failure reports, (including one complete failure and a partial failure on the other engine) resulted in five deaths. By far the single greatest cause of total accidents other than pilot error in the Seneca series of aircraft during the period reviewed were mechanical failures. One trim runaway (which was fatal) several brake failures, a jammed elevator and a nose bowl separating in flight (which was also fatal) were sighted. There was also one report of an in flight fire just after departure. The largest number of mechanical failures goes to the

landing gear. Six incidents were attributed to landing gear malfunction/failure. Because Seneca's are used as training aircraft it isn't surprising to find four accidents where the instructor let the student go a little to far. Weather played a role in four incidents, two reports sighted wind as a primary cause, two others sighted ice. Night or night IMC were sighted in three other incidents. There were two in flight break up's during the period studied, one attributed to night, the other to night and icing.

The pilot error accidents include one pilot who ditched his plane in the ocean after an engine failure with seven people on board (the Seneca is certified for six). Another flew while medically impaired, another who took off with the trim set in the full nose down position, and another who beat his aircraft against the runway after a bounced landing and a botched recovery. Several pilots chose to land downwind and ran off the end of the runway, and two pilots flew their aircraft into the ground in controlled flight (CFIT) during missed instrument approaches. One pilot came up short when he ducked below the glide slope on final, and two ran their aircraft completely out of fuel resulting in off airport landings. There was one mid-air (fatal) and one pilot hit a horse on the runway while landing at night.

PA-31 Navajo

There were 90 accidents reported in the PA-31 series aircraft from between period 1/1/96 through 11/1/00, 26 of which were fatal. Of interesting note is the fact that 45, fully 1/2 of the incidents occurred while operating under Part 135.

Pilot error while shooting instrument approaches was the single largest cause of fatal incidents reported. One pilot killed himself while shooting an approach while under the influence, and two other died in a stall spin accident during a Part 135 training flight. The single largest loss of life occurred when a sightseeing plane, with 10 on board crashed on a volcano in Hawaii. The NTSB could find no cause for the crash. There were two reported incidents of pilots having a heart attack in flight. Both those incidents were fatal.

There were a large number of mechanical failures attributed to PA-31's as well. They include several engine fires, attributed to a faulty fuel pump. One hydraulic pump failure lead to landing gear failure, as well several reports of jammed/in-op landing gear doors. The cowl departed two aircraft, striking the tail.

There were 9 examples of pilots running the aircraft out of fuel. And in three of those incidents, in-op fuel gauges were reported as the primary cause of fuel exhaustion. There were two incidents of mis-fueling, where the aircraft was fueled with Jet A. There were four confirmed reports of pilots landing with the gear up.   There was one report of a pilot who hit a moose on landing roll. Eight loss of control reports during either take off or landing following an engine failure.

So, what can we derive from this? What land mines can we clearly see with the benefit of hindsight? How can we learn from the misfortunes of other fellow aviators? As you read NTSB reports you can't help but say to yourself "I would never do that!" I can assure you that the pilot who hit the Moose while landing at night or the pilot who took off with the trim in the full nose down position had similar thoughts. Lets analyze what we can from these reports and discuss practical ways to reduce your risk while operating a multiengine aircraft.

When you look at the Cessna 310, you can see that 16 of the 82 reported incidents were fatal or about 19.5%. It is interesting to note the number of incidents, which were fuel system related. It would seem obvious that this model has an operational problem with the design of the fuel system since so many pilots simply ran it out of fuel. The 414 had a fatal accident rate during the study period of 23%. The 421 had 10 fatal accidents out of 30 reported accidents, or 30%. Now remember, this is total fatal accidents, not total fatalities. It stands to reason that the aircraft with more seats has the statistical ability to kill more people per incident. Neither the 414 nor 421 displayed any accident trends that could be attributed to poor or inadequate design like those found on the fuel system of the 310.

But when you eliminate the unavoidable accidents such as hitting an animal on the runway, and the accidents that are the results of purely bad judgment, like flying with a known mechanical problem, what are you left with? In all aircraft, single and multiengine with retractable landing gear, there are numerous reports of pilots who failed to put the gear down for landing, and in some cases, retracted the gear after landing. A few years back I read an insurance industry study, which drew an amazing coincidence between the numbers of gear up landings in twin engine aircraft in relation to the number of hours left before TBO on the engines of the accident aircraft. The inference being that maybe, just maybe, some of those gear up's were not accidental. But the fact is, at least in piston twins that we as general aviation pilots are simply not prepared to handle the situation when we loose an engine. The statistics don't lie. More incidents / accidents occur where the primary or incidental cause is an engine failure.

Yes, there are often other mitigating circumstances such as weather, ice, night, and even some things a pilot can't control such as mechanical failures and mis-fueling. But on a whole, the accident reports are filled with stories of pilots who simply were not up to the task when the feces hit the fan. Many of the pilots who experienced engine failure in flight made it all the way to the airport, only to loose it on the landing roll out. Some lost control with an engine out at night on an instrument approach, and others while negotiating their way through ice. These are what I would call compound events, which led to an accident.

So even though you are rated and current, are you competent to handle a compound emergency? The statistics say "NO". What is the answer? Well interesting enough, the answer may come from the insurance industry. A few years back the aviation insurance underwriters essentially made it impossible to obtain insurance on turbine powered singles and twins without taking type specific initial training in an approved school. It didn't matter if you were a 20,000 ex-airline captain or a 1,200 civilian trained CFI. Even for aircraft that do not require a type rating, you would receive initial training as if it was for a type rating.

A few years back I took initial training for a C-90 King Air. This King Air has 8 seats and a cruise speed of about 225 kts. Fairly close to many of the piston powered cousins we are flying. Additionally, the C-90 has a gross weight of under 12,500 lbs, so no type rating is required. That program included 20 hours of classroom, and 20 hours of simulator training. In the class we would review each and every system on the aircraft, from electrical, through pressurization, in finite detail. By the end of the program, I could tell you which major sub systems are powered by which electrical buss and what to expect if that buss were to go off line. Then we would go into the simulator and actually practice compound emergencies.

Was it realistic? You bet! In fact, I find simulators are slightly more sensitive then the real airplane. Master the sim and it's easy in the plane.

There is simply no way to practice some of these failure modes in an airplane for real. Over pressurization, under pressurization, partial and complete gear failures, electrical failures of every kind, anti-ice / de-ice system failures....all can be safely recreated in the simulator.

It doesn't end there. Every year we are required to do a recurrent training course, which is about 20 hours, again split between class and simulator. Does it work? Well, turbine singles and twins have an accident rate about half of piston singles, or about 4 times less than piston twins.

The good news is that this type of training is available for most piston twins. There are several schools such as Sim-Com in Orlando Florida, Flight-Safety and Simuflite, which offer programs for piston twins. While this training isn't cheap, it clearly improves your chances of successfully surviving a compound failure situation, which could lead to a fatal accident.

Of course, nothing is a complete substitute for time in the aircraft. Go out and find a qualified instructor. If you are unhappy with your first choice, find another. When you do find one you like, hang onto him or her as if you life depended on it, because in a sense, it just might. Confess your weaknesses to them. You know you better than anyone. Then let them take you through situations, which will build your experience and confidence in handling the things that make you uncomfortable.

Not comfortable at night? Have them take you flying at night. Not happy in ice? Go with the instructor into a known ice situation (assuming the aircraft is certified and it is all working of course). Set your self up a regular program. Several hours every 90 days, every six months, what ever it takes. I'm a CFIIMEI/ATP and fly three hundred plus hours a year, much of it IFR. Yet I still take an IPC every six months even though I have met and exceeded the minimum requirement for currency by a long shot.

The point is that you should learn what works in your equipment from people who know, and then hone your personal skill and knowledge of the machine and keep it sharp! 135 pilots take a check ride every 6 months. Just because your operating under part 91 doesn't mean you don't need the training, it just means you don't have to demonstrate competency. (There are a lot of things you can do legally under part 91, like take off in zero/zero weather for example, but I would have to question the judgment of a pilot who would do it).

Clearly, it takes a larger commitment to training and currency in a twin then in a single, and if you wish to operate safely, you need to be realistic about your ability to meet that commitment. Don't kid yourself! The statistics don't lie!

For the last few pages we have been talking extensively about safety in multiengine aircraft. You've read about it, you understand that the twin can offer a quantum leap in capability and performance over most of it's single engine brethren, if your willing to put in the time to become and stay a competent pilot.

What does it cost to operate a twin? How about buying one?

If you look on page 126, I discuss the cost of operation for piston engine aircraft using a multiple of fuel burn formula. For the most part, that will work fairly well for most piston-powered airplanes.

The turbine world is an entirely different deal. Without a

doubt, the most difficult question to answer is "What does it cost me per hour to operate this type of airplane?"

For what would appear to be a straightforward question, there is only one accurate answer, and that is "It depends."

I want to give you the broad general overview of those considerations, your operating options and their associated costs as well as some real world costs for operating a light and medium turboprop, and a light jet.

The first consideration in your cost analysis thought process is the acquisition costs. We are discussing used aircraft here. Airframe and engine times as well as other life limited components and required inspection times have a tremendous effect on the value of a particular aircraft. Avionics also affect the value of an aircraft, though less so than engine condition. A word of caution here; If you have no experience with aircraft ownership, or even if you have experience in the piston powered world of aviation, know this; Buying a turbine powered aircraft is no place for beginners. A mistake here can cost in excess of a hundred thousand dollars down the road. On used turbine aircraft, engine condition dictates the value of the aircraft in a disproportionate ratio to the airframe. It really doesn't matter how long it has been since the last "Hot Section" because it only takes one "hot start" to cook a guide vane or warp a compressor blade, and replacement/repair costs start in the tens of thousands of dollars. Time remaining on other life limited components will have a direct bearing on the cost of the aircraft.

In the light jet market, compliance with R.V.S.M. (reduced vertical separation minimums) is an expensive, one time fix. You will need to consider other available options at overhaul time such as the M.O.R.E™ program. This program, which stands for Maintenance On Reliable Engines, is a component life extension program, which can extend the useable life of certain turboprop engines from as little as 3,000 hours to up to 8,000 hours.

Realize that aircraft trade like commodities. After 9-11, turbine aircraft prices appeared cheap (relative to accepted "blue book" valuations). Two years after the 9-11 event, approximately 25% of the turboprop fleet was looking for a new home. Lots of aircraft with few buyers yielded lower prices.

The same thing happened in 1991 during the first Gulf war. The economy looked questionable, and aircraft were cheap.

Timing also has an effect on your tax situation. Simply stated, under the right set of circumstances, you can qualify for a large depreciation expense in the first year on a used aircraft. When you make the purchase will affect how much depreciation you qualify for in the current tax year.

The second major consideration is insurance. If you are a small company, or have a seven figure net worth, and you are comfortable with between one and five million dollars in liability insurance, you have a set of options available to you that if you're a publicly held company, or have a high net worth simply won't work. If your legal council is thinking ten to twenty five million in insurance coverage, you will need to take a different avenue.

The next issue is your mission profile. The mission profile defines what you need to accomplish with the aircraft and crew. The aircraft that satisfies 80% or so of your mission profile is the one you want to own.

All of this leads to operational considerations. As an aircraft owner, you have basically three options when it comes to operation of your aircraft. One option is to buy an aircraft and hire a pilot or pilots if the aircraft requires a crew or fly it yourself.

Option two is to hire an aircraft management company to operate your aircraft for you under FAR Part 91, but not for compensation or for hire.

The last option is to purchase an aircraft for the express purpose of air commerce, and lease it back to a qualified aircraft charter company that will operate the aircraft under FAR Part 135 (Commonly called charter). Placing the aircraft on charter will allow you to take advantage of the tax incentives. Charter generates cash flow, thus reducing ownership costs. It is kind of like leasing back your Skyhawk to the flight school.

Let's take a look at these three options and see how they affect operational costs.

Fuel, insurance, hangar and even maintenance costs are directly affected by the management options you choose. Fuel makes up about 30% of the direct operating cost of a flight, and is a major consideration. Our little charter company bought more than 300,000 gallons of Jet –A last year. That earned us some very significant fuel price discounts. Fuel providers give operators volume discounts not normally available to individual operators. The same is true of hangar space and insurance.

Lets take a look at three aircraft, each representative of their class of plane. The King Air 90, is a typical of a light turboprop, a King Air 200, which represents a medium turboprop, and a Citation 1, which is representative of an entry-level jet.

For the comparison purposes we have used acquisition costs for aircraft in the middle of the price range for the model. Hangar space costs are a national average for that size aircraft. Financing costs are included but <u>none</u> of the available tax advantages are reflected nor is any revenue from charter considered in these figures. All the per-hour operating costs are sans crew, and in 2004 dollars. Please read the assumptions at the bottom of the chart for a better definition of the expenses.

	King Air C90	King Air 200	Citation 501
Acquisition cost	$650,000	$950,000	$1,200,000
Hanger	$7200.	$10,200	$10,200
Insurance	$13,825.	$18,300	$19,500
Computer Fees (7)	$3,500	$3,500	$4,600
GPS Updates	$600	$600	$600
Fuel	$261.25	$282.50	$337.50
Maintenance	$129.86	$175.38	$214.48
Parts	$70.72	$152.91	$120.71
Reserves	$77.14	$98.33	$108.15
Per Hour	$547	$925	$1,098

†The values shown are compiled from Business and Commercial Aviation, using Aviation Research Group research.

ASSUMPTIONS:
1. Aircraft flown 300 hours per year
2. Owner flown
3. 100% business use.
4. Aircraft financed with 80% LTV, 7% interest for ten years
5. Liability Insurance limits $1,000,000.
6. These costs do include most of the normal cost of operation of the aircraft although all items are not listed to conserve space.
7. Maintenance on turbine aircraft is generally kept on a computer program like Cescom or Maintenance Director.

If you will need to hire a crew, it going to cost about $200 per hour for a single pilot crew, and about $350 per hour for a two-man crew whether you go managed or hire your own. (This is in 2004 dollars). It is our opinion that you will have more flexibility and less liability in a managed aircraft situation then trying to create your own flight department.

If an aircraft were used less than half the time for business use, it would certainly be wise to consider placing it in charter service to obtain the maximum benefit from the tax code.

There is another compelling reason to consider a managed or charter type program. That is to protect yourself from yourself. Many successful business types think that aviation is a business like theirs, and are bottom line driven. Their business experience has taught them that the penalty for a bad decision is that it costs money. In aviation, the penalty for bad judgment can be fatal. Just as you would not lie on the operating table and tell a surgeon how to perform an operation, you don't want to put yourself in a situation where you are sitting in the right seat of your corporate aircraft telling the pilot how to conduct the flight.

All of the factors to be considered for such a decision are far more complex then can be discussed here. My best advise is to seek the consultation of a professional who can provide some guidance on the decisions that are outside your personal paradigm.

So.....You want to fly a King Air ???

Turning procedure turn inbound the weather is at minimums on the NDB 7 approach. To make matters worse, we are covered in ice, the left sub buss breaker has blown taking out most of the panel lights as well as the captain's side instruments. As I roll out level inbound the right engine fire warning light illuminates and I begin executing the shutdown procedure for the right engine. Once that generator is off line I'll be counting on the battery to put the gear and flaps down. "500 above and one minute to go" says the co-pilot as I prepare to manually crank down the gear. "MDA and 30 seconds to go" he says as I hold the gear to the very last minute. We break out right at minimums, "I have the runway," he says, and I begin cranking. There is no gear actuator or indicator with a failed left buss so I pull the power lever for the dead engine back to listen for the horn. No horn... the gear must be down. I look up, see the lights, and deploy the flaps. Suddenly the plane begins to roll violently to the left. "Split flaps" I call and immediately retract the flaps regaining some amount of control, and plant the airplane on the runway. At least I kept it on the black part. True story? Well.... Yes, and no. Just another day in the simulator at Simcom / Pan-Am Flight Academy in Orlando.

I went there to get King Air qualified. A good buddy of mine bought one of these twin turbine, cabin class twins and needed someone to fly with him until he could be insured for single pilot operation on his own. The insurance companies know that these type specific programs reduce their liability exposure and require this type of training annually, even for the most experienced pilots.   So although the FAA doesn't require a type rating for the King Air C90, the insurance company does.

I did not have a lot of turbine experience so I was looking forward to a taste of that world.

My first impression of the school was how carefully they built the program to cater to owner pilots. But being an A&P definitely helped as we got into discussions on pressurization, electrical systems, propeller and engine controls, de-ice, and anti-ice as well as environmental systems. The King Air, like all aircraft in its class, are complex machines. About as much time is spent in the classroom studying the systems as is spent in the simulator playing with them. Spend two hours in the classroom discussing electrical failures, and then spend two hours in the simulator experiencing them.

Although Simcom's simulators are not motion simulators, the quality of the graphics is good enough to fool your inner ear into thinking you're pitching up or down or banking left or right. I had the opportunity to look at Simcom's new PC-12 simulator and it had improved graphics over what was in the King Air and King Air 200 sims I got to fly. But while they don't have motion, they do come with a sim instructor guaranteed to challenge even the most seasoned veteran.

My partner in class was a 19,000-hour, ex Marine Corp fighter pilot, airline type with more ratings than would fill the page and he got his fair share of excitement as well. In fact, I learned a lot sitting right seat for him when he flew the sim.

Hot starts, hung starts, run-away engines, over pressurization, under pressurization, anti-ice, de-ice failures, gear failures, electrical failures of all kinds can be experienced, explained and learned to be dealt with without the risk of damage to a real aircraft, not to mention the expense.

What impressed me the most was how easy the King Air is to handle when everything is going right, and what a handful it is when things begin to go wrong. Unlike its piston cousins, where you could memorize a simple checklist, in the King Air, checklists are a must, and there are plenty of them.

Some of the nice features in the King Air that us piston guys haven't seen before are features like Auto Feather. Lose an engine on take-off? The prop feathers itself. All you need do is stay in control until a safe altitude is reached and then you can proceed with the shutdown and secure checklist. There is plenty of power, even on one engine. A fully loaded King Air C90 like ours will climb in excess of 500 feet per minute on one engine at gross. The later models with the bigger engines do even better. Another neat feature is the Beta mode for the propeller.

Essentially, it allows you to put the props in reverse to slow down on a short field or on a slick runway with poor braking. You can even back up should you have to.

One thing no one will accuse a King Air of is being too fast. The relatively low operational speeds make for an easy transition for any current competent multiengine piston pilot.

At about 220 knots it's not much faster then some of the high performance piston aircraft. But it makes up for that in pure comfort. It has a large, comfortable, air-conditioned cabin.

The turbine engines produce plenty of bleed air for pressurization, and our King Air will make a sea level cabin at 10,000 feet and a 7,000-ft. cabin at 24,000 feet. Although the airplane is certified to 30,000 feet, it is impractical to take it there, and requires a different stand-by oxygen system than is currently installed.

Additionally, according to accident statistics, the King Air is by far, the safest of the class, having the lowest fatal accident rate of about a dozen types listed in its class.

Having survived a week of simulator training, and receiving my certificate (suitable for framing my instructor pointed out), it was time to fly the real deal.

Start up was just like the simulator, except in the sim you don't smell the Jet-A. Real life lesson number one....close the storm windows before starting the engines. Taxi was interesting as well, using Beta to slow and turn the plane.

Running through the check lists prior to take off were identical to the sim, the only difference being the radios, which slowed me down looking for the correct knob, button and switch.

On the take off roll I couldn't help but be impressed by the acceleration (pun intended). As I was pushed back into the seat the airplane went through Vmc plus 5 and off we went climbing at close to 2,500 feet per minute, which is typical at the weight we were at. We might have used 1500 feet of runway. Having flown Pitts type airplanes before, I wasn't behind the airplane as we went through one thousand feet, lowering the nose and reducing the power and props. Watching the temps and torques took some time just getting used to where to look and what you are looking at, but enough like the simulator so that in a few minutes that task became subliminal.

The next thing I discovered is what every King Air pilot already knows.....what a nice flying airplane the King Air is. Light on roll, slightly heavy on pitch the King Air is stable as a table, and very easy to fly. Minimal rudder pressure is required to coordinate a turn and there is no mistaking the Beechcraft "feel" to the controls.   It felt a lot like my Twin Bonanza, which shouldn't surprise me since it is essentially the same wing. The large trim wheel on the center pedestal makes trimming the airplane precisely, an unconscious thought. In fact, for most of the time I preferred the manual trim to the electric trim on the yoke.

Shooting the ILS was far easier than in the sim as well, as the aircraft is far more stable in pitch and roll than the simulator. Gear at the marker gave me 600 feet per minute down and that was exactly what I needed to stay on the pipe.

I did a missed approach just to see how busy things could get. One nice feature on this autopilot is a "go around" button, which commands the bars on the flight director to a predetermined pitch-up attitude. Just put the little red triangle in the yellow bars on the flight director and you can't get it wrong.

Next, I wanted to play with the engine out handling characteristics. Pulling one engine all the way back, the airplane obediently yawed into the "dead" engine. No surprise there. What was surprising was that even with the prop windmilling, there was plenty of power available to climb the plane without feathering the prop. Keeping the nose straight with rudder is do-able, but for anything more than a few minutes you are going to want to use the trim.

Now it was time to land the beast. My 19,000 hour ATP buddy offered to show me how short you could land a King Air, so I sat back and watched. A red line approach to touchdown, followed by judicious application of Beta resulted in ground roll of less than 1000 feet....without touching the brakes.

So what do I think? I think that a pilot considering a late model Seneca, Baron, Bonanza or Mooney may want to consider this; The King Air we are flying is a 1971 model. It has new everything, including paint, interior, and radios. Avionics include color radar, Stormscope™, KFC-250 Flight Director / Auto-Pilot with altitude preselect and yaw damper, dual HSIs (one of them a new Sandel™ EFIS type), radar altimeter, RMI, even an IFR approved GPS. We had props overhauled, and the engines have been upgraded on the MORE™ program to extend service life another 4,000 hours. All of that, plus turbine engine reliability, and virtually all weather capability for under half a million dollars. Before you say "whoa......that's a lot of money" realize that is less than a new piston twin anything, and about the same as a new high performance piston single. The difference is a quantum leap in performance and comfort not to mention capability. Obviously, it only matters if your mission profile calls for something like that, but judging from the sales of Pilatus P-12s, there is definitely a demand. It also comes at the price of more frequent and more intense training. A twin turbine aircraft is simply no place for the weekend warrior. Not to mention higher maintenance costs.

As for me, I told my buddy when he is ready to move up to an Eclipse Jet and he's done with the King Air, I want to be first in line to buy it!

As for the sim training; well, five years after I took it, I was giving dual in a King Air and experienced a torque runaway. In that failure mode, the engine goes to full power and you can't control it. The only option is to shut it down.

I had done them in the simulator, but it is not something you can simulate in an aircraft. At first I thought the other engine had failed, but then the training kicked in. I identified the errant engine and made the shutdown decision, all inside of a few seconds. We landed without further incident and other then a slightly shaken student, none the worse for wear.

I ultimately did buy that King Air and I am still as impressed with it as the day I first flew it.

Learning to Fly the Citation Jet

The date on the back of the faded snapshot hanging on my office wall read "MARCH 1966". Two little boys and who appears to be their mother, posing in front of a Luscome 8A. Not a particularly unique photo except that it captured the exact moment in time I knew I wanted to be a jet pilot. I was one of those little boys, and on that warm spring morning on a long gone grass strip on the south shore of Long Island, New York, that little airplane ride started me on a 36 year journey that ended in the front seat of a Citation.

I don't remember much about being seven years old, but I remember that day. I remember the sheer panic on my mom's face as I pleaded for permission to go for the ride offered to me, begging in a way only a seven year old can. I was scared to death, but drawn to that machine as if somehow I knew I just had to go for that ride. I remember the moment we left the grass runway, the loud cacophony of sound inside that cockpit, and the sounds and smell of the spring air rushing in past the sliding window. I remember the pilot, a friend of my mom, flew me over my house and pointed out my dad's car parked in the driveway. I could see my friends playing baseball on the street in front of my house who had no clue I was directly overhead. I remember returning to the airport and touching down on the grass, taxiing up to my anxious mother, holding onto my little brother. Hopping from the plane, I thanked the man for the ride, stopping in front of the aircraft just long enough to take a picture, that picture.

From that day on, I dreamed about being a pilot. No, a jet pilot. I could use my childhood imagination to turn virtually anything into a jet-powered flying machine. Give me a ball-point pen, and in 30 seconds I could imagine it a jet plane.

When we went to visit grandma, who lived in the flight path to Kennedy Airport, I would sit for hours and watch the planes whiz by on their way from where ever.

Like all kids my age, I built model airplanes. Flying models, plastic models, metal, plastic, wood, even paper, I really didn't care.

At age 10, my mom took my brother and me on the first of many trips to Europe, Asia and the Middle East. My first airline trip was on a TWA Boeing 707 from Kennedy to Madrid, Spain. At the airport I bought a copy of Janes World Airline Atlas, and by the end of my trip I could tell you the difference between a DeHavilland Comet and a Vickers Viscount.

High School brought the reality of making a living into full view. But the only way to the airlines was through the military and in 1976, the U.S. armed forces were in the middle of the biggest down- sizing since the Second World War. The only way you could get a seat in a military cockpit was to graduate from one of the service academies, and the only way you were going to go to one of those was with a congressional appointment. It didn't take long for me to realize that that wasn't going to happen. So, plan B.

A degree in graphic arts technology assured that I wouldn't starve to death. A good job in New York had me earning big money (for a kid my age in 1980). One day a buddy called me up with a hot stock tip. "Integrated Barter," he said, "It's a quarter but it is going to a dollar.... Call your stock broker!" Yeah, like I had one. By the time I got one on the phone the stock was 35 cents but I bought it anyway. Yes, it went to a dollar. Actually a dollar and a quarter. I sold the stock and took the money to the local flight school in Morristown, New Jersey and started flight training.

Fast-forward twenty years. Seventy aircraft, and a fist full of ratings later, I find myself closing in on retiring from the business that has afforded me the ability to own my own aircraft, and fly what I want. I am seriously looking at getting into the turbine charter business with a man who has more time upside down then I have logged total time, and I hold an ATP rating. The first flight with him in our first turbo-prop clearly illustrated (to me) just how far I had to go before I was the type of pilot I wanted to be.

The leap to the turbo-prop world from the piston world was dramatic. The requisite training for part 135 operations for turbo-prop aircraft is the same as that for a type rating, though the weight of the aircraft (less than 12,500 lbs) precludes the necessity to have a type rating. I struggled to get my mind around jet engine theory, ITT's, P3 bleed air, N1 speeds, over torque and over-temps, hot starts and hung starts.

Operating in the higher altitudes posed its own set of transitions from the piston world. Ice, even in summer, 100 kts plus headwinds, and much higher operational speeds challenged my flying abilities. But most of all, in the world of fly for pay, the customers want to go. They don't care about weather, ATC delays or the fight you had with your wife. When they get to the plane, it is show time.

But the business was good to us. Two years into the deal, we were operating five turbo props, including a DeHavilland Dash-6 as well as two piston twins. Last month we flew well over one hundred revenue-producing hours. It was time to consider the next step, and that step was into the world of the jet.

After a good long hard look at our local market, which is dominated by older Lear Jets, my partner and I decided on the Citation. Slower than the Lear, and retailing out at a rate higher per hour than the Lear some questioned our decision. We could buy three Lear 24s for what a single clean Citation 501 costs. But many of our customers are charter brokers and operators and they felt that many of their turboprop customers might step up to the Citation on long trips because it is comfortable and quiet compared the cramped and noisy Lears. One sitting in a Lear cockpit convinced me. There was no way I was spending more money than I spent on my house for an aircraft I couldn't fit in comfortably. Decision made. Now all I needed to do was learn how to fly one.

We had been doing our flight training with SIMCOM in Orlando, Florida. They had recently merged with Pan Am Flight Academy and moved into a new building just east of the Orlando International Airport. Part of the deal brought a full motion simulator for the Citation over to SIMCOM so we decided to do our training there. My partner was already Citation typed, so all he needed was recurrent training. I needed the entire type rating.

Day one went well until they issued me two books for my training course. One was a Citation technical manual, the other the Citation flight manual. Both are about the size of the Manhattan Yellow Pages. For laughs I weighed my book bag....38 lbs.

The type-rating course consists of a full 40 hour week of ground school, followed by 20 hours of simulator training and the check ride.

I learned right away that the transition from turbine to jet was not nearly as big a leap as was the transition from piston to turbo-prop. Many of the systems on the Citation were similar to what I had on the turbo-props I had been flying, so unlike the transition from the piston world I had a frame of reference from which I could relate to the new plane. I could not image trying to go from piston to Citation. I'm sure it could be done, but it would take a better pilot than me.

In the 40 hours of classroom that happen before you ever get near the simulator, every system in the aircraft is reviewed in great detail. Some systems, like the air cycle machine used for air conditioning were new to me. Anti skid braking, auto engine starting, and oh yes, no prop levers were all new to me. In a lot of ways, the Citation is a simpler aircraft than some of the turbo-props to operate, with many tasks automated.

Another interesting aspect of the Citation training is the CPT, or cockpit procedure training and CRM, cockpit resource management. This is where you learn to fly as a crew. The turbo-props we fly were all single pilot but the Citation, (some of which can be operated single pilot under Part 91), is definitely a crew served (read two pilot) aircraft in the Part 135 world. All of a sudden, flying has become a team sport. Success and failure in the left seat, depends on the pilot in the right seat. In fact, the right seat work load is greater than that for the pilot in command. Strict adherence to check lists, plus the use of oral call outs to initiate specific actions just have to happen properly or both pilots get behind the airplane and the flight looks ugly. Talking on the radio, setting up the nav radios, briefing the pilot on the approach, are all the responsibility of the first officer.

In our class of six, two pilots were transitioning straight from the world of piston twins, one other had some heavy turboprop time. Another was a former Air Force fighter pilot and also a former Simcom instructor who just could not stay retired. I partnered up with a guy who flew turboprops for one of the big fractional programs and was going back to corporate flying. His flight experience was similar to mine and we seemed to work well on the pre-simulator procedure training.

Finally, after seven days of book pounding, a written test; to no one's surprise, each of us knew the airplane cold. Our first shot in "the box" came that afternoon. The simulator at Simcom is a class "C" training device. It is full motion, night only. That means the F.A.A. considers it is an airplane. It has a tail number and everything. It stands two stories tall and you get in by walking across a catwalk connecting the sim to the second story of the building. You are even briefed on how to egress, should the thing break down or should we have a power failure or something. The entire machine pitches up and down, left and right and is in sync with a windshield graphic, all designed to fool your inner ear into believing you are actually flying. It takes about 10 seconds to forget you are firmly anchored on the ground. So realistic is this device, you must wear your seatbelts and shoulder harnesses, because if you were to crash, you could be thrown from your seat!

Engine start checklist complete it was time to go "flying". Immediately you are impressed by the way you get pushed back into your seat. The aircraft accelerates so quickly that you are through V1 and ready to fly long before your mind says "rotate". The two engine rate of climb, especially when light in weight is a huge departure from anything that is piston powered. I've flown Pitts and Extras and they are truly impressive compared to any Cessna or Piper you have ever flown, but the jet is flat out awesome.

We blow through our assigned altitude, both I and my co-pilot are still behind the airplane. Level off at a low altitude and you need to be careful not to exceed Vne, which is 262 knots.

On our first ride in the sim the instructor doesn't fail anything. It will be the only time in the next 20 hours that everything in the aircraft works. During the next 20 hours we will experience explosive decompression, complete with sound effects, electrical fires, bleed air failures, pressurization failures, engine failures of every type, single engine approaches and single engine go-arounds. (A concept totally foreign to a piston twin pilot). Low visibility takeoffs, even single engine takeoffs are discussed and practiced. One other reality hits home right about now. Your basic instrument flying skills had better be first rate before you get here or you are done.

By the fourth sim day, day eleven, I am exhausted. This type of training is intense. Sixty hours of instruction in 14 days is a very brisk pace. Three days and a check ride to go! It also happened to be my 44th birthday. When class was out I went back to my R.V. which I kept at a campground just a few miles from the school and sat by the lake on a comfortable spring evening. So close, I'm almost there.....just a few days left to go.

Next day, more emergencies, rapid descent, (in excess of 6,000 ft per minute. At least that is where the VSI is pinned), un-commanded thrust reverser deployment in flight, flap failures, landing gear failures, brake failures, sometimes all on the same approach! Next, unrecoverable wind shear. Even if you do it all right, you only have a 60% chance of surviving. Each and every approach is to minimums, at night, and with a crosswind.

Day 12. We start in the classroom learning the differences between Citation types. The rating, for a CE 550, is for a Citation II, but allows you to fly most of the 500 series Citations. The differences between the 500, 501, 525 and 550 are subtle but important. Different operating speeds, limitations and procedures depending on the specific model and in most cases, which serial number aircraft you are flying. Then, more simulator time.

The last day of simulator training includes a mock check ride; complete with steep turns, a stall series and a circling approach to minimums at, of all places, Kennedy Airport in New York.

At this point, my training partner and I are in the groove. We have developed a comprehensive pre take-off briefing, we have worked on our call outs, nailed our memory items, and we are ahead of the aircraft not behind it. Both of us are flying as one. When one misses a checklist item, the other chimes in. We are doing it, we are flying as a crew! For me at least, this has been the most difficult part of the training. For twenty years, I have been a single pilot. It is hard to break twenty years of flying habits, but if you want to fly jets, you must learn to fly as crew.

Check ride minus one day. Today we do L.O.F.T.training. L.O.F.T stands for Line Oriented Flight Training. It's a simulation or a "normal" day of flying in a Citation. Route clearances, departures procedures, and standard terminal arrival routes, step down fixes, crossing altitudes and speed restrictions. Speed restrictions! An aspect of aviation most piston drivers never worry about. My training partner and I ace it. We have some extra time so we elect to review steep turns and stall series, plus emergency descent. As has been the routine all week, one system after another failed on the approach. Touchdown was without flaps, speed brakes, wheel brakes or thrust reversers. Fortunately, the runway was long. Of course the approaches were all to minimums. We handled it. Bring on the check ride!

Check ride minus 12 hours. I can't sleep. Procedures, call outs, memory items are all running through my head. I turn on the T.V. and watch the news. The weatherman tells me that the planets will all be visible in a straight line at 2 a.m. It hasn't happened in thousands of years and won't happen again in my lifetime. It IS 1:50 a.m., close enough. I go outside and peer into the night sky. Sure enough, I can see several planets in more or less a straight line vertically, from just above the horizon through about 12 o'clock.

I marvel at this silent nature show for a few minutes. Too bad it happens so late at night, otherwise I would call my son and tell him to go outside and check it out. The distraction relaxes me enough to get to sleep. Good thing, tomorrow's a big day.

Check ride day. The examiner explains the routine. No tricks, just what we have been trained for. Good, we are ready.

A little overanxious, we rush through the first approach, a tricky VOR 4R to JFK, with a circle to land on 31 right. In the real world, no sane pilot would accept that clearance, but this isn't the real world. Next, we fly a hand-flown ILS approach to minimums for runway 9L at MIA. On departure from there, we climb up to 10,000 feet for air work. 45 degree left and right turns, followed by a clean stall, a departure stall and a dirty stall, all to A.T.P. standards.

Then, the localizer 30 approach to Miami International. On the way in, an emergency descent to 2,000 feet. We lose the flaps and one engine. The trim runs away and the auto pilot fails.

We are a busy crew. Touchdown uses most of the runway. As the 1,000 foot remaining centerline lights whiz by, we deploy the emergency braking system blowing the tires on roll out. We stayed on the centerline.

On departure from Miami, we suffer an engine failure and fire on the take-off roll. Our next attempt results in an engine failure at V1, so we take our problem into the air. An air restart is fruitless. We are sent to an intersection to hold. Suddenly the weather gets better, and we are cleared for a visual approach to runway 12. A single engine approach and landing completes my "half" of the checkride. We switch seats and do it all again for my training partner.

Two hours later, I am sitting in the briefing room watching my left hand sign my name on a temporary airmen's certificate with my name on it. Along with all the things it used to say, it now says CE-500. I am too tired to appreciate what has just happened.

"Leaving one seven thousand for flight level three nine oh". The words come out of my mouth like I have said it a thousand times, but this is my first time above thirty thousand feet, and my first time in command of a jet.

Sitting in the cockpit I find the presence of mind to savor the moment. The smooth and relative quiet of the jet engines are a quantum leap beyond the turboprop, which I had felt was a quantum leap beyond the piston aircraft I have flown all my life. It is everything I had hoped for and a few things I never would have guessed. I would not have guessed that those large windows, which come down almost to the pilot's elbows and provide amazing visibility, also heat the cockpit to uncomfortable levels while the passengers are freezing. Additionally, I now pack sun block 35 in my flight case after receiving a nasty sunburn....on one arm.

Up ahead, a hazardous weather advisory for thunderstorms with tops to 35,000 feet actually doesn't affect us. We slide effortlessly over the weather without so much as a bump, the passengers never even knew.

So thirty-six years and a few weeks after that moment in time frozen by the camera I finally got to be a jet pilot. Was it worth the wait? Was it worth the time energy and effort? It scares me to figure the actual cost, as just the type rating itself was more than ten thousand dollars. The time? I don't know, nearly three thousand flight hours and as many invested in ground studies. That's roughly three years of a full time job. I personally believe that had it not been for an amazing confluence of circumstances there would be no way I would be here.

Oh sure, I've heard all the jokes about Citations. "Slowtations", the only jet susceptible to a bird strike from the rear, when discussing near-jet speeds, the Citation is the near-jet they are talking about. I guess the definition of "speed" is relative. But for me it has been an amazing journey and I wouldn't trade it for the world.

Tail Wheel Tips

In Wolfgang Langerwiesche's classic book about flying, *Stick and Rudder,* he writes, "In trying to understand how airplane and pilot behave during a landing, we had best begin at the end." As applied to flying a tail wheel equipped aircraft there is no better advice.

The tail wheel itself evolved to accommodate hard surface runways. Originally, aircraft were equipped with tailskids. They were operated off of aerodromes. These round grass fields assured take offs and landings were always into the prevailing wind. When hard surfaced runways began to appear the tail wheel was the simple solution for directional control of the aircraft while on the ground.

So why all the mystique about flying tail wheel aircraft? Part of it comes from the fact that the F.A.A. requires an endorsement simply to act as pilot in command of one. Remember, at one time they were called conventional gear aircraft, but for the last 40 plus years, the nose wheel equipped airplane has been the configuration of choice because it is inherently more directionally stable on the ground thus easier to teach primary flight in.

Following Langerwiesche's advice, let's start at the end and discuss landings in a tail wheel equipped aircraft.

First, if you have never actually flown a tail wheel aircraft, forget all the hangar talk that you have heard. There is nothing inherently difficult in learning to fly a tail wheel aircraft, except perhaps finding an instructor who can teach you properly. In the 13 years I have been teaching tail wheel, I have taught every type of pilot from a zero hour spouse of an owner to a 24,000 hour L-1011 driver. The key elements of being a good tail wheel pilot remain the same.

#1) **Three point landings;** Most tail wheel aircraft prefer to be landed in a three point, full stall configuration. Yes, there are models that don't like that, but by and large, most general aviation tail wheel equipped aircraft land best in a three point, full stall landing. The reason for that is minimal forward speed. Energy is exponential to speed, so the slower the touchdown speed, the more manageable the landing will be. I tell my students to hold it off, hold it off some more, don't let it land, keep pulling the stick back all the way to the stop. You should hit the stop before touchdown. The most common mistake I see is the impatience of the pilot. Trying to force the aircraft onto the runway before it stalls only insures that if you bounce, you will go flying again. Be patient. Wait for the plane to quit flying. This way, when you touch down, the aircraft is done flying. If the aircraft is equipped with a stall horn or light, it should be on. The second most common mistake is to release backpressure on the stick or yoke after touchdown. You must keep the tail nailed to the runway with full aft stick in order to maintain tail wheel effectiveness.

#2) **Use of brakes;** Most modern general aviation tail wheel aircraft have very effective brakes. They should never be used for directional control while landing. Again, there are exceptions to that rule, but in general, keep off the brakes except to taxi.

Using differential braking for directional control is, in my opinion the second easiest way to induce a ground loop. I have my students keep their heels on the floor and away from the brakes until we have slowed to taxi speed.

#3) **Proper crosswind technique;** In much the same way that GPS has created an entire generation of pilots who can't maintain situational awareness using VOR's much less NDB's, so has the nose wheel equipped trainer created a generation of pilots with poor crosswind technique. While the average flight school trainer might let you get away with being five, even ten degrees or more off of runway heading at touchdown, the average tail wheel equipped aircraft is simply not forgiving of anything less than perfect form. Whether you use the forward slip or the crab/kick method of wind correction doesn't matter, only that the aircraft is aligned with the runway at touchdown and whatever control inputs that are required to keep it that way are employed.

#4) Positioning of control surfaces; Another forgotten art that takes on great significance when operating a tail dragger is the proper positioning of flight controls while on the ground. Whether taxiing or on the rollout, you must remember to position the controls so that you climb into and dive away from the relative wind. Failure to continue to move the controls into the wind on roll out can force the tail wheel to slide or skid sideways, leading to loss of directional control or worse, a ground loop.

#5) Wheel landings; Perhaps the most misunderstood techniques of landing a tail wheel aircraft is that of a wheel landing. This technique, which is employed when wind conditions are such that a full stall landing may require more rudder authority than the aircraft can produce at stall speed carries its own inherent risk.

In a wheel landing, you are flying the airplane onto the runway at a speed well above stall, in a relatively flat attitude, and touching down only on the main wheels. The stick or yoke is pushed slightly forward at touchdown to keep the mains firmly planted on the runway. The higher speed equals more airflow over the tail, thus more control authority. The tail is held in the high position, off of the runway until the stick or yoke is full forward against the stop, then allowed to fall to the runway. Upon contact, the yoke is immediately pulled full aft, nailing the tail to the ground. The risk of course, is that with all the extra speed, the aircraft is more than willing to go flying again should you bounce, or your timing is off. Over application of forward stick pressure can force the main gear to spread thus momentarily changing the landing gear geometry and the way the airplane tracks on the ground. In some designs, the aircraft runs out of rudder before the tail comes down, thus necessitating the use of differential braking for directional control, which as I said earlier can be dangerous.

The bottom line is flying a tail dragger requires a certain amount of training and experience, and a slightly higher level of competency and attention to technique than their nose wheel equipped cousins. The rewards for taking the time to master the tail dragger will impact your flight performance in whatever you fly, whether it is a rented trainer or a heavy jet. It opens up another world of pre-war designs for you to explore and they too become candidates for ownership. It will sharpen your stick and rudder skills and bring them to a level that may be unachievable in a tricycle gear aircraft. But most of all, it allows you bragging rights to say, " I'm a tail dragger pilot!"

The More Things Change....

It's hard for me to fathom, but I am rapidly approaching my 25th year as an airman. Last year I celebrated twenty years of aircraft ownership. Yesterday, I took my plane up for a ride to nowhere in particular. As I pulled the plane out of the hangar the neighbors kid, a young man of about seventeen, walked over to admire my bird. Within a minute I asked him to join me, which he did eagerly. As we flew over the rapidly disappearing farm fields of western Palm Beach County, he asked me questions about learning to fly and getting a pilot's license. It started me thinking of my early days of flight training in Morristown, New Jersey just 35 miles from the busiest airspace on the planet. Interestingly enough, the cost of flight training is about the same as what I had paid more than 20 years ago. Back in 1982, Cessna had a 40-hour program in a C152 for $1,995.00. But nobody did it in 40 hours and almost everyone upgraded to the 172. I believe I spent a little over thirty three hundred dollars learning to fly. Divide that by the 61 flight hours in my logbook at the time of my private pilot checkride and that works out to about $55.00 per hour. That is about what you're going to pay at a small rural airport flight school these days. There is little else in this world I can think of that costs the same or nearly the same as it did in 1982. By any definition, that would make flight training a bargain. But if that were the case, why are flight schools hurting, student starts declining and the all-important number, pilot certificates issued been in a steady downward slide since the early eighties? I guess it depends on whom you ask.

In my opinion, general aviation has done a really poor job of marketing itself. In the last 20 years, there are far more activities vying for the limited disposable income of the average consumer. The boating industry, all but devastated by the demise of the investment tax credit in the early eighties, re-invented itself in the form of personal watercraft. In doing so it reached a much broader audience. Radical sports like para-chute jumping, rock climbing and white water rafting have all developed into full fledged industries to cater to the adren-alin junkies of the late baby boom generation as well as the next great wave of consumers, the Generation "X" ers.

In some respects, the very nature of General Aviation itself has changed. Twenty years ago experimental aircraft were the lunatic fringe of aviation. Now, they are considered mainstream. Doctors, lawyers and businessmen, once consid-ered the mainstay of general aviation patronage are now readily found among the ranks of the experimental crowd.

Give the experimental movement some credit for keeping the flame of interest burning during the late eighties when piston powered general aviation aircraft production fell to just a few hundred aircraft per year from around 18,000 annually in the late seventies.

Some things haven't changed. If you haven't been to a flight school lately, take a look the next time you're at the air-port. Most small operations are flying 25-year-old aircraft, and some of them look every day of it. Now when you're 21 years old and you have the burning desire to fly maybe that's o.k. When you're in your early forties and heavily insured you look at that plane and think "Hell No!".

As a student pilot, in my first 40 hours of flying I experi-enced total electrical failure at night, vacuum failure on my first solo cross country trip, and partial power loss due to a broken carburetor venturi, which allowed the engine to develop all of about 1,600 rpm. The deal breaker for me was having a mixture control handle come off in my hand while airborne. The frayed cable end still attached, I had no way of knowing if the engine would continue to run.

When I teach primary, I will only teach in a new or late model aircraft. Not because I think old airplanes are unsafe or ugly. No, it is because the student will.

Technology just hasn't kept up with expectations when it comes to aviation. I don't think I have seen a new car without air-conditioning in 20 years. But very few new planes come with it. Even if you wanted to pay for it, on top of the "more than a new house" price for the machine, it simply isn't available. So how many middle age businessmen and women really want to strap into an airplane on a 90° day and shoot an hour's worth of touch and goes?

When I was a student pilot I went through a series of young instructors, all on their way to flying careers elsewhere. They were poor teachers, and couldn't care less because they were logging their time on their way somewhere else. I nearly walked away from aviation one afternoon when an instructor put us in a spin, then told me to recover, without ever briefing me in advance on HOW exactly I should do that or even that we were going to do that.

When we got back on the ground I explained to him that he would find it difficult to fly anything while in a body cast, and he should probably not do that to me again. He actually though the whole episode funny. A talk with the flight school operator revealed the problem. Instructors of any caliber were hard to find and the operator was reluctant to chase this one off.

Well guess what, good instructors are still hard to find. Pilots still use the instructor's position as a way to log flight time, and flight schools are still reluctant to do much about it.

If it wasn't for the acquaintance of a former U.S. Navy F-4 Phantom pilot whom I had met through non-aviation business activities, who took the time to finish my private pilot training, I might have walked away. I promised him that one day I would get my instructor's rating, and I would spend as much time and patience as it took to work my students through their training. Six years later I got my CFI. My double I, and MEI after that. I've never had a student fail a written or flight test.

I have never had a student bust a check ride and as far as I know, none of my students or former students have ever been violated, broken an airplane or hurt themselves while flying.

I have always felt that it was my personal mission to prevent a pilot from walking away from aviation and to make a new pilot whenever possible. Fortunately, I don't have to feed my family or myself on what I make flying. The old joke about how to make a million dollars in aviation (start with two million) is the painful truth.

Flying itself has become a lot more complicated. TRSA's, and ARSA's and airport traffic areas have given way to class A, B, C and D airspace. Flight service stations, once found everywhere, are now down to just a few. When once you could walk into the FSS and see the briefer, talk with him or her, look at the charts and raw data, now they are disembodied voices on the automated phone line or a computer generated report. We now use METAR and TAF as opposed to SA's. It is still in code, just in a different language.

We now have altitude-encoding transponders in almost everything, and need to, in order to operate in or even near the nation's busiest airspace. On the positive side, electronics have revolutionized navigation, communication and engine management. While we pilots would like to think it was all done for us, you can find the same cool moving maps, fuel flow gauges and MFD's on a mid priced boat. There are millions of boats in the U.S. and 185,000 privately owned aircraft.

Since September 11th, you need to be a Philadelphia lawyer just to understand the "Notam du Jour", and if you get it wrong it is entirely possible that an F-16 will show up on your wingtip pulling a banner that says "Deposit License Here".

Would I do it all over again? I don't know. If I could be twenty years old again, maybe I would. I learned to fly in the first place because it had been something I had dreamed of since I was five or six years old. I had the fire for it for sure. If that were not the case I would have walked away long ago. At this point in my life I still feel that fire. I will still go flying to nowhere in particular and still take the opportunity to introduce a fresh face to the wonders of aviation. Good, bad or otherwise, General Aviation is what General Aviation is, take it or leave it. The unfortunate reality is that more and more of us are choosing the latter.

To be perfectly fair to the General Aviation industry, it has been burdened with tremendous product liability issues, that even huge industries like healthcare can't contend with, so it is not surprising to see that few are willing to risk it all in such a litigious environment for such a limited return on investment.

But American enterprise has amazing resiliency. When Home Depot came to town everyone including the Sunday talk show talking heads thought it was the end for the small independent hardware store owner. Yet, they managed to reinvent themselves and that segment of the hardware / home-improvement market has become the fastest growing, most profitable section of that industry.

And so I believe it shall be in general aviation. On September 1st 2004, the new Sport Pilot rules become effective, creating a new class of airmen. This rating essentially fixes all the problems inherent in the failed Recreational Pilots rating.

The new rating does have many limitations over a private pilots license. A holder of a Sport Pilot Certificate is limited to day VFR. In order to fly in any class B, C or D airspace, the Sport Pilot Holder must have received dual instruction in that specific class B, C or D airspace in advance, and have a logbook endorsement attesting to that instruction for that specific airspace. Other limitations include no flights above 10,000 msl, no flights to foreign countries, (without advanced permission of that country), no flights with flight or surface visibilities of less than 3 miles and no flight without visual ground contact.

Medical qualifications are also simplified. If you have a current driver's license issued from one of the 50 United States, you can fly as a Sport Pilot, within the limits imposed on the driver's license. So for example, if you are required to wear corrective lenses to drive your car, you will need to wear them to fly. Your FAA issued third class medical will work too.

More limiting is the type of equipment that a Sport Pilot Certificate holder can fly. Most of the aircraft affected are currently non-certificated (read experimental) aircraft, though the development of this new rating is expected to generate more "factory" built aircraft that meet the new criteria.

That criterion features a maximum gross weight of 1,320 lbs, (1420 lbs for a float equipped aircraft), with a maximum stall speed of 45 knots and a level flight speed not to exceed 120 knots. It can only have two seats.

Qualifying aircraft will be limited to a single engine, which cannot be a turbine engine; the aircraft must have a fixed or ground adjustable propeller and fixed landing gear. It cannot be pressurized.

What is interesting is that they can be manufactured, ready to fly, without compliance with FAR part 23-certification compliance. Manufactured aircraft will be required to follow ASTM standards for materials and procedures. That means less expensive to purchase, less expensive to maintain, factory built aircraft. How the insurance industry will accept these aircraft has yet to be observed.

A few certificated aircraft can fall under the Sport Pilot Certificate rule. The Piper J-3 Cub, as well as the J-4 and PA-11 are all 1320 lbs or less. So too, are the Aeronca 7AC series, the Ercoupe 415 C, the Interstate S1A, Porterfield FP-65, and the Taylorcraft BC-65 series.

The good news is that these new rules should generate more interest and participation in general aviation. The bad news is that we should expect a certain amount of teething pains associated with the new rules.

So what of the next twenty five years? I'm willing to offer some predictions.

The micro-jet will become a reality. Whether it is Eclipse or Saffire or the Cessna Mustang I can't say, but they are coming. Ushered in by a wave of new power plant development,

I suspect small, affordable turboprop engines are not far behind.

I'll also predict that leaded aviation fuel will go away. As I understand it, it represents about 3/10ths of 1 percent of all the motor fuels refined in this country, so I don't expect tears from the oil refiners. But for us high performance piston engine owners, I predict improved engine technology like FADEC to become more common place as well as piston diesel engines that can run on Jet-A.

I think you will find in-cockpit data-link technology in every cockpit, similar to what the Capstone project has both promised and delivered in Alaska, though I'm not sure if it will be provided by the government or come from the private sector.

I'll bet you will see stricter regulatory compliance issues for commercial operators, conforming more or less, with what we see internationally. Private pilots may be affected too. Even as I write this, there is a movement to restrict VFR at night, or to require an instrument rating and an IFR clearance to operate at night. That is how it is in most of the rest of the world.

I predict that entrepreneurial business people everywhere will ultimately embrace General Aviation, as flying commercially becomes more time consuming and less cost effective in the post 9-11 reality of the hub and spoke commercial airline transportation system. NASA'S Small Airport Transportation System concept is the right idea in the hands of the wrong salesmen.

I run an aircraft charter business for a living and I can tell you that business is good. It is a hard business, with thin margins and is without a doubt the most regulated industry on the face of the earth, but there are people like me who love what they do and will put up with the bureaucracy to be able to continue to do it.

People still need to go places, and cost has become less important than time.

I think it is general aviation's time.

I've seen a lot of general aviation in the last twenty-five years. Some of it I saw coming, some of it, I never would have guessed. With any luck at all, I'll see a lot more in the next twenty-five years, and I'll still be writing about it.

About the Author

Michael Leighton began flying in 1980. By 1990 he held his private, commercial, instrument, multiengine and Certified Flight Instructors ratings. By 1993, he had added the Instrument and Multiengine Instructor ratings as well.

A Single engine Seaplane rating, an ATP rating, and a type rating in Citation 500 series jets, along with an Airframe and Power Plant Mechanics license all followed.

Along the way, he has owned more than seventy different aircraft ranging from a very rare Travel-Air 12Q biplane to a King- Air 200.

He now operates a FAR Part 135, Air-carrier Company located in South Florida.

His aviation stories have been published in various aviation magazines continuously since 1995.

You can reach him via e-mail at av8tor0414@aol.com